T0073567

Unearthing Fermi's Geophysics

Unearthing Fermi's Geophysics

Based on Enrico Fermi's Geophysics Lectures of 1941

Gino Segrè and John Stack

University of Chicago Press

Chicago and London

The University of Chicago Press, Chicago 60637
The University of Chicago Press, Ltd., London

Published 2021
Printed in the United States of America

30 29 28 27 26 25 24 23 22 21   1 2 3 4 5

ISBN-13: 978-0-226-80514-6 (cloth)
ISBN-13: 978-0-226-80528-3 (e-book)
DOI: https://doi.org/10.7208/chicago/9780226805283.001.0001

Library of Congress Cataloging-in-Publication Data

Names: Segrè, Gino, author. | Stack, John (Physicist), author. | Fermi, Enrico, 1901–1954.
Title: Unearthing Fermi's geophysics / Gino Segrè and John Stack.
Description: Chicago ; London : The University of Chicago Press, 2021. | Includes bibliographical references and index.
Identifiers: LCCN 2021013461 | ISBN 9780226805146 (cloth) | ISBN 9780226805283 (ebook)
Subjects: LCSH: Geophysics.
Classification: LCC QC806 .S44 2021 | DDC 550—dc23
LC record available at https://lccn.loc.gov/2021013461

♾ This paper meets the requirements of ANSI/NISO Z39.48-1992 (Permanence of Paper).

Contents

Preface **xi**

Acknowledgments **xv**

1 Introduction **1**

1.1 Fermi's Notes 1

1.2 Plan of the Book 3

2 Gravity and the Earth **7**

2.1 The Earth's Gravity 7

2.2 Earth's Gravitational Potential 10

2.3 Pressure at the Center of the Earth 19

2.4 The Moments of Inertia of the Earth 21

2.5 Motions of the Earth's Axis 24

2.6 Equinoxes and Solstices 24

2.7 Precession of the Equinoxes 25

2.8 Free Rotational Motion 29

3 Heat and the Atmosphere **35**

3.1 First Law of Thermodynamics 36

3.2 State Variables 36

3.3 The Second Law of Thermodynamics 38

3.4 Ideal Gas 41

3.5 Chemical Potential 51
3.6 The Boundary between Phases 54
3.7 Moist Air 58
3.8 Atmospheric Composition and Temperature . 63
3.9 Temperature vs. Altitude 63
3.10 The Upper Atmosphere 65

4 Atmospheric Loss **69**
4.1 Maxwell Distribution 69
4.2 Escape Velocity 70
4.3 Lifetime of Atmosphere 75
4.4 Magnitude of Particle Losses 76
4.5 Helium 78
4.6 Speed of Sound 79

5 Liquid Drop Physics **83**
5.1 Vapor Pressure and Raindrops 83
5.2 Liquid Drops Affect Vapor Pressure 86
5.3 Properties of Falling Raindrops 88

6 Coriolis Effects **93**
6.1 The Coriolis Force 93
6.2 Coriolis Effects in the Atmosphere 97
6.3 Hurricanes 103

7 Radiation at the Surface **107**
7.1 Surface of the Earth 107
7.2 Crust of the Earth 108
7.3 Composition of the Oceans 109
7.4 Surface Temperature Equilibrium 109
7.5 Black Body Radiation 110
7.6 Planck's Formula 114
7.7 Special Cases 116

8 Ocean Properties **119**

8.1 Heat Current 120
8.2 The Heat Equation 121
8.3 Diurnal Temperature Variation 123
8.4 Temperature of Seawater 126
8.5 Ocean Circulation 128

9 Gravity Waves **131**
9.1 Dispersion Relations 131
9.2 Euler's Equation 132
9.3 Small Amplitude Surface Waves 134

10 Tide Physics **143**
10.1 Effects of the Moon 143
10.2 The Tidal Potential 147
10.3 Water-Covered Earth 147
10.4 Time Dependence of Tides 150
10.5 Tides on Lakes and Seas 154

11 Earthquake Properties **157**

12 Seismic Waves **163**
12.1 Elementary Seismographs 163
12.2 Seismic Waves 165
12.3 Strain and the Strain Tensor 165
12.4 Stress and the Stress Tensor 168
12.5 Hooke's Law 171
12.6 Young's Modulus and Poisson's Ratio 171
12.7 Stress and Strain 173
12.8 Newton's Law in a Solid 175
12.9 The Equation of Motion for Displacement . . 176
12.10 Longitudinal and Transverse 177
12.11 Snell's Law for Elastic Waves 180

13 Surface Seismic Waves **187**
13.1 Surface Waves 187

13.2 Oscillations of the Earth 189
13.3 Breather Mode 189

14 Radioactivity **197**
14.1 Historical Note 197
14.2 The Sixth Washington Conference 198
14.3 The Earth's Density 200
14.4 Estimates of the Age of the Earth 201
14.5 Radioactive Decay Equation 202
14.6 Radioactivity within the Earth 202
14.7 Uranium Decays 206
14.8 Radioactive Dating 207

15 Heat Flow in the Earth **211**
15.1 Surface Heat Variations 213
15.2 Heat at the Earth's Surface 214
15.3 Scales in Heat Conduction 215
15.4 Heat Equilibrium 217

16 Earth Magnetism **223**
16.1 Units 223
16.2 Magnetic Moments 224
16.3 Magnetic Field of the Earth 225
16.4 Earth Magnetization 227
16.5 Rotation of Charges 229
16.6 Currents inside the Earth 230
16.7 Magnetosphere 233
16.8 Magnetic Storms 233
16.9 Magnetic Potential Expansion 234

17 Atmospheric Electricity **237**
17.1 Overview 237
17.2 Properties of the Electrosphere 238
17.3 Water Drops in an Electric Field 247

18 Atmospheric Plasma **251**
18.1 Ions in the High Atmosphere 251
18.2 Atmospheric Radio Waves 252
18.3 Other Effects 256
18.4 Double Refraction 256

Afterword **261**

Appendix A Earth Days **263**

Appendix B Torque on the Earth **267**

Appendix C Rayleigh Waves **271**

Bibliography **279**

Index **284**

Preface

The geophysics course from which this book is derived was offered by the great physicist Enrico Fermi in 1941 in the physics department at Columbia University. It met every week for a single hour-and-a-half session. It was primarily intended as a senior-level elective for science and engineering undergraduates. Enrollment presupposed a mathematical preparation in partial differential equations and vector calculus and one in physics that included courses in mechanics, thermodynamics, and at least some basic notions of electricity and magnetism.

The same prerequisites for a geophysics course hold true today. Some subjects, such as the theory of elasticity, are not contained in the usual undergraduate physics curriculum, nor are they presented in the notes, but, are necessary for understanding Fermi's discussion of seismic phenomena, so we have included the necessary material in our text. In a few cases we have gone beyond the discussion Fermi provided: the inclusion of the equations governing the precession of the equinoxes is one example, a more complete discussion of the thermodynamics of perfect gases is another, and a fuller elaboration of tidal motion is a third. In a few other instances we have briefly introduced topics that we felt Fermi might have suggested his students consider as challenging extensions of the material he had presented. The thermodynamics of hurricanes, breather mode oscillations of the Earth, and water drops in electrical fields are examples of such topics.

This book is not meant to include all the material necessary for an introductory course in geophysics, but we believe geophysicists will find it interesting for historical reasons and that any scientist or engineer with an adequate physics background will profit from a study of its contents. Many of the topics it treats, such as thermodynamic and electrical properties of the atmosphere, will generally not be contained in a conventional introductory geophysics course. The resulting text is a concise overview that readers will find intriguing as an introduction to this broad, fascinating subject. In remaining faithful to the original scope of the notes, we have made no attempt to bring its contents up to date by the inclusion of topics unknown or insufficiently clarified in 1941, such as plate tectonics or the geodynamo theory of the Earth's magnetic field.

The eighteen chapters follow the order of the original presentation of that same number of topics, but the reader may choose to follow another sequence. Each chapter is essentially self-contained, so any single one can be consulted as a separate entity. In every case, following the lead Fermi provided, the presentation is made as simple as possible. A few technical appendices provide additional details.

In the intervening decades a switch has been made to SI units from the centimeter-gram-second (cgs) units common at the time Fermi presented these lectures. We have thought the reader would find it easier if we used SI units as well, though admittedly it has been useful at some points to present cgs results, e.g., density in grams per cubic centimeter, in order to discuss the numerical estimates. The same is true for electromagnetism, where at the time of Fermi's lectures, electrostatic (esu) units were common. In esu, the unit of magnetic field is the gauss (G), whereas in SI, the unit of magnetic field is the tesla (T). For the magnetic fields near the Earth, the gauss is the more natural unit ($1\text{G} = 10^{-4}\text{T}$), so in the following text, magnetic fields are usually stated in gauss.

We have also reproduced in this text reproductions of a few of the thirty-eight pages (not all complete pages) of Fermi's notes, as found in the Regenstein Library Fermi Archives. This is done in order to give the reader an impression of how they appeared to us. They have been placed immediately preceding the chapter in which they appear. Readers desiring to see the full manuscript of Fermi's notes on geophysics may access them at **https://press.uchicago.edu/sites/fermis_geophysics/index.html.**

Acknowledgments

G. S. would like to thank Bettina Hoerlin for her patience and support, Philip Nelson at the University of Pennsylvania for discussions, and Diane Harper and Barbara Gilbert at the Regenstein Library of the University of Chicago for their assistance with the Fermi notes. We both would like to thank Victoria Outerbridge for expert help transforming Fermi 's hand-written notes into the images included in the book. In addition, we both would like to thank Christie Henry, now at Princeton University Press, for the initial enthusiasm about the project, as well as Alan Thomas and Michaela Luckey for continued editorial support at the University of Chicago Press. We would most particularly wish to thank Joseph Calamia for his support and assistance in all matters, large and small, leading to the completion of the book.

CHAPTER 1

Introduction

1.1 Fermi's Notes

While studying material on Enrico Fermi at the University of Chicago Regenstein Library to aid in the preparation of a biography of Enrico Fermi one of us (Gino Segrè) was writing in collaboration with his wife, Bettina Hoerlin, found a folder labeled "Geophysics, Columbia (1941)." It appeared to be a set of sketchy personal notes for a course on geophysics that Fermi had taught in the physics department at Columbia University in 1939, 1940, and 1941. Though the contents of the folder were surely clear to Fermi, Segrè felt the material they contained would not be evident without the benefit of a detailed explanation. He was intrigued by them as well as being surprised that Fermi would have chosen to teach an introductory course on geophysics as his initial course offering upon arriving in America. That puzzle was later solved by discovering that in 1928, young Fermi, already recognized as a genius and freshly chosen as Italy's first professor of theoretical physics, had taught Italy's initial offering in quantum physics *and* an additional course in geophysics: incidentally, this had been a regular offering in Rome's physics department at that time. In other words, Fermi had arrived in America as an experienced instructor of geophysics, and, always interested in all branches of physics, he had doubtless continued to study the subject.

This led Segrè to envision the project of presenting to a scientific public both the schematic notes and an explanation of the material in them, aiming to achieve this goal at the level of an advanced undergraduate physics course. Concerned about the task at hand, Segrè was fortunate to enlist an old friend to collaborate on the project: John Stack, a physics professor at the University of Illinois in Urbana. In addition, having been a graduate student of Geoffrey Chew, one of Fermi's first American students, Stack was also a "physics grandson" of Fermi.

Though geophysics has undergone great advances since the time of the Fermi lectures, many of the subjects and techniques Fermi used are not out of date, so we felt such a monograph could be instructive for students as well as being of historical interest. We therefore decided to describe in detail the topics he introduced and how he dealt with them because Fermi, one of the titans of twentieth-century physics, had certain personal characteristics that made any set of lectures by him especially interesting.

One of these is that he was a famously lucid lecturer, admired by the physics community for his unparalleled ability to present complicated subjects in simple terms. Second, he was known in a wider community for both his interest and ability to estimate the magnitude of any physical phenomenon, a capacity that he encouraged his co-workers and students to develop. His skill in such endeavors became so well known that such puzzles are now simply known as "Fermi questions." This facility is on full display in these lectures, where every topic covered includes quantitative results on the size of the effect under consideration.

1.2 Plan of the Book

Geophysics, as taught by Fermi, is a somewhat idiosyncratic set of lectures. The course includes items about the Earth that we would expect in any introductory geophysics course: its motion within the solar system, the shape of the Earth and its composition, heat in its interior, elasticity and seismic waves, geomagnetism, and atmospheric physics.

Other topics might not normally fall within the purview of geophysics lectures at present. Ocean currents and tides would appear in an oceanography course, the fall of raindrops in one on fluid mechanics, details of radioactivity as a case study in nuclear physics, and many of the thermodynamics calculations in a set of lectures on that subject.

Several key new developments in geophysics, plate tectonics being an outstanding example, are not mentioned in the notes. This is understandable because although the roots of this subject go back to the work on continental drift by the explorer and meteorologist Alfred Wegener in 1915, the theory was not widely accepted until the 1960s, twenty years after these lectures. Other subjects, such as geomagnetism, have undergone such radical changes that much of the material in the lectures on that subject is mainly of historical interest.

It is also sad to see the absence in the lectures of the important changes ushered in by the use of electronic computers and the novel study of nonlinear dynamics, for these are areas in which Fermi was a true and important pioneer in the years immediately preceding his untimely death in 1954. He certainly would have eagerly followed the advances in modeling of the atmosphere, the oceans, and the Earth's interior that have depended on powerful computers. Computational geophysics was essentially nonexistent in 1941 but now is a mature field.

Except in minor instances we have not attempted to bring these notes up to date, for doing so would distort what is after all the perception of them that Fermi had in 1941. Our aim in presenting this book has been to follow closely his notes, discussing the topics he raised in the manner and in the order he adopted for his presentation. But on occasion, particularly when we have felt a fuller discussion would help the reader, we have tried to present one. In a few instances we have found it irresistible to add an extension of a topic Fermi was discussing. Our main concern has been to attempt to elucidate the frequently telegraphic style of the notes. This has meant, inter alia, to provide derivations and explanations of equations that Fermi used as jumping-off points.

We hope to have produced a set of notes similar to what an attentive student might have prepared for the course Fermi taught in 1941. We have also tried to incorporate its many numerical estimates, realizing that some of them may not have been meant for inclusion in the lectures but were rather a reflection of his well-known habit of continually checking formulas numerically. Except for the few cases when we thought it would cause confusion, the notation we have adopted is the same as the one Fermi used.

Though a reader may wish to examine the topics in an order other than the one we have presented, by, e.g., joining together all the sections on the atmosphere, and separately all those of the interior of the Earth, we felt compelled to remain faithful to the selection made by Fermi in his notes, though we are well aware that this may not be the actual order in which he presented the material during his lectures. But we have tried insofar as it was possible, to be consistent with the spirit of those lectures

Since our version of his lectures has been prepared without the benefit of oversight by Fermi, we are naturally open to the accusation that we have misread or misinterpreted what he intended to say. We have tried to remedy this pitfall by making available his personal notes on the material, as we found them in the Chicago archives. At the very least, this volume will have the merit of allowing an interested reader to see how a physics genius with a strong interest in geophysics approached teaching this fascinating subject and what topics he deemed appropriate to include at the time he taught a course on the subject.

We are sensitive to the charge that Fermi would not have liked to have his lectures see the light of day in their present format. There are certain discussions in them that are incomplete and even a few numerical errors, obviously the result of these being preliminary notes. But we have felt that the interest in seeing the choice of topics Fermi made and his approach to them has justified the preparation of the present volume. We only hope the reader will share our judgment and forgive our errors.

Eratosthenes about 200 y b.C. (first measurement of radius) ①

Principle of geometrical & astronomical measurements

Results Equatorial ½ axis $a = 6,378,388$ m

Polar ½ axis $b = 6,356,911$ m

$$\frac{a-b}{a} = \frac{1}{297}$$

~~Mass of~~ Volume of the earth Surface

$V = 1.082 \times 10^{27}$ cm^3 5.1×10^{18} cm^2

Mass of the earth

from $g = 980 = \frac{GM}{R^2}$ $M = \frac{(6.36 \times 10^8)^2 \times 980}{6.67 \times 10^{-8}} = 5.95 \times 10^{27}$

$G = 6.67 \times 10^{-8}$ $\rho = \frac{M}{V} = 5.5$ (5.52)

Rotation 1 solar day = 86400 sec

1 sidereal day = 86164.09 sec = T

Centrifugal force at equator $\frac{4\pi^2 a}{T^2} = 3.46$

$g_{equator} = 978.052$

$g_{pole} = 983.21$ $\Big\}$ $g = 970\,(1 + 0.00531 \cos^2\vartheta)$

Flattening for centered mass

$$\frac{a-b}{a} \cong \frac{2\pi^2 a}{gT^2} = \frac{1}{2}\,\frac{1}{g}\,\frac{4\pi^2 a}{T^2} = \frac{3.46}{2g} = \frac{1}{570}$$

Flattening for homogeneous mass

Parameters of the Earth

CHAPTER 2

Gravity and Precession of the Earth

2.1 The Earth's Gravity

Erastothenes of Cyrene, a third-century BCE. Greek, is commonly regarded as the father of modern geography. He not only regarded the Earth as a sphere but succeeded in calculating its circumference by measuring the difference in the length of the shadows cast at noon in localities a considerable distance apart in a north-south direction. He was also the first to calculate the tilt of the Earth's axis. That is where matters stood until the seventeenth century when observers noted that the Earth was not a perfect sphere. Isaac Newton put forth the notion that because of the centrifugal force the Earth was in fact better approximated as an oblate flattened ellipsoid. The next significant advance was Alexis Clairaut's publication in 1743 of *Theorie de la figure de la terre.* This contains a theorem on the variation of the gravitational acceleration over the Earth. We return to this in Sec. (2.2) below.

Table 2.1: Gravitational data used by Fermi

a	6,378,388 m
b	6,356,911 m
T_{sol}	86,400 s
T_{sid}	86,164.09 s
G	$6.67 \times 10^{-8}\,\mathrm{cm^3/g\,s^2}$
g_e	$978.052\,\mathrm{cm/s^2}$
g_p	$983.21\,\mathrm{cm/s^2}$

Fermi takes up the story by considering the variations of the acceleration due to gravity over the Earth's surface. Table (2.1) contains the values of various physical quantities used by Fermi. In Table (2.1), a is the equatorial semi-axis of the Earth, b is the polar semi-axis, T_{sol} is the length of a solar day, T_{sid} is the length of a sidereal day, G is Newton's gravitational constant, g_e is the acceleration due to gravity at the equator (including centrifugal effects), and g_p is the acceleration due to gravity at the poles. A solar day can be defined to a very good approximation using a simple sundial. With an approximate error of only 30 s/year, a solar day is the time it takes the sundial's shadow to make a first return to a specific position. The figure of $T_{sol} = 86,400\,\mathrm{s}$ is 24 h/day $\times$ 60 min/h $\times$ 60 s/min. A sidereal day is approximately 4 min shorter than a solar day. It is measured by considering the Earth's rotational motion with respect to the far away or "fixed" stars rather than with respect to the Sun. In most of our subsequent discussion, the distinction between T_{sol} and T_{sid} will not play a role and will be ignored. However, for the interested reader, more detail on the distinction between T_{sol} and T_{sid} is provided in Appendix A.

The quantity known as the "flattening" is

$$\frac{a-b}{a} = f, \tag{2.1}$$

where f is a commonly used notation. Fermi writes

$$\frac{a-b}{a} = \frac{1}{297}, \tag{2.2}$$

a value which follows from the data in the table. For the volume of the Earth, he gives

$$V = 1.082 \times 10^{27}\,\text{cm}^3. \tag{2.3}$$

This is close to the value obtained by assuming the Earth is an ellipsoid of revolution. For that case, we have

$$V = \frac{4}{3}\pi a^2 b = 1.083 \times 10^{27}\text{cm}^3.$$

Fermi then goes on to an elementary calculation of the mass of the Earth assuming the Earth is a sphere of radius $R_e = a$, with acceleration due to gravity given by $g = 980\,\text{cm}^2/\text{s}$. Newton's law of gravity states

$$g = \frac{GM_e}{R_e^2}, \tag{2.4}$$

so

$$M_e = \frac{gR_e^2}{G} = \frac{(6.36 \times 10^8)^2 \times 980}{6.67 \times 10^{-8}} = 5.94 \times 10^{27}\text{g}, \tag{2.5}$$

comparable to the current best value of 5.97×10^{27}g. For the density of the Earth, Fermi gives

$$\rho = \frac{M}{V} = 5.5\,\text{g/cm}^3,$$

similar to the current value of $5.51\,\text{g/cm}^3$. It should be noted that the numerical values given by Fermi are typically correct to "slide rule accuracy," the slide rule famously being Fermi's personal hand calculator. This occasionally leads to discrepancies in the second decimal place. From here on, we will not comment on this sort of discrepancy.

2.2 The Earth's Gravitational Potential

Geographical Coordinates Suppose we set up a coordinate system with the origin at the center of mass of the Earth, and the positive z axis pointing north along the Earth's axis of symmetry Any point can be specified relative to this coordinate system by giving its spherical coordinates; r, θ, ϕ, where r is the distance of the point from the center of mass, and θ and ϕ are the usual polar and azimuthal angles. For a point on the surface of the Earth, θ is often called the "co-latitude," the latitude being $|\theta - 90°|$, plus the designation N or S, depending on whether θ is less than or greater than 90°.[1] The azimuthal angle ϕ is closely related to the longitude. The longitude of a point on the Earth's surface is the angle along a circle of constant latitude, measured either east or west from the Prime Meridian, which runs through Greenwich, a borough of London. Following Fermi, we will use the spherical coordinates θ and ϕ.

Nonspherical Effects Having defined our coordinates, let us turn to the gravitational potential due to the masses in the Earth. We denote the Earth's gravitational potential as V_g. There are, of course, gravitational effects near the Earth that arise from the Moon, the Sun, and other planets and stars, for example the precession of the equinoxes and the tides. These are discussed in Sec. (2.5) and Chap. (10), respectively. However, with respect to the gravitational acceleration near the Earth, the effects of other astronomical bodies can be neglected compared to those that come from the Earth itself.

The gravitational potential due to the Earth has a value at

[1] For example, the latitude of Chicago is 41.9°N, while that of Buenos Aires is 34.6°S.

any point inside, outside, or on the Earth. For a small mass μ which is acted upon by the Earth's gravity, the potential energy is μV_g. Since gravity is an attractive force, we expect $V_g < 0$, with $-\mu V_g$ being the work required to move the mass μ out of the Earth's field to infinity.

Although the gravitational potential at points inside the Earth is certainly of interest, we will restrict the present discussion to points outside the Earth or on its surface. In this region, satellite measurements have determined the Earth's gravitational effects with great accuracy. To start off, suppose the Earth were perfectly spherical. In that case, V_g is very simple;

$$V_g = V_g^{(0)} = -\frac{GM_e}{r}. \tag{2.6}$$

It is one of the remarkable features of Newton's law of gravity that for a spherical body the gravitational potential obeys Eq. (2.6) at any point outside or on the surface of the body. But the Earth is not perfectly spherical. As briefly mentioned at the beginning of Sec. (2.1), it is slightly oblate, with a larger radius at the equator than at the poles. (See Table (2.1).) It is clear that the zeroth-order term $V_g^{(0)}$ needs to be augmented by a correction term, which must depend on the polar angle θ, in addition to r.

There is a systematic procedure for finding angle-dependent corrections to $V_g^{(0)}$. This is to make use of the fact that, outside of the Earth's mass distribution, the complete gravitational potential satisfies Laplace's equation,

$$\nabla^2 V_g = 0. \tag{2.7}$$

Eq. (2.7) is directly analogous to the statement that away from any charges, the electric potential satisfies Laplace's equation.

If the Earth's oblateness is the only nonspherical feature that needs to be included, a correction term of the form

$$\frac{3\cos^2\theta - 1}{r^3} \tag{2.8}$$

is sufficient. It is easy to check that the formula of Eq. (2.8) satisfies the Laplace equation. Adding a term of this form to $V_g^{(0)}$, we now may write

$$\begin{aligned} V_g &= V_g^{(0)} + V_g^{(2)} + \ldots \\ &= -\frac{GM_e}{r}[1 - \frac{1}{2}J_2(\frac{a}{r})^2(3\cos^2\theta - 1) + \ldots] \end{aligned} \tag{2.9}$$

In Eq. (2.9), a is the Earth's equatorial radius, and J_2 is a dimensionless parameter whose modern value is $J_2 = 1.0826 \times 10^{-3}$. It is fair to ask, what property of the Earth's mass distribution determines the value of J_2? As will be explained in Sec. (2.4), the answer is that J_2 is determined by the moments of inertia of the Earth. The terms denoted by $+\ldots$ in Eq. (2.9) represent a sum over higher-order corrections that can account for more detailed angular dependence than $V_g^{(2)}$. The general form of the lth term in the series is

$$\frac{N_l(\theta, \phi)}{r^{l+1}}, \tag{2.10}$$

where $l = 3, 4, \ldots$, and N_l is a polynomial (Legendre function) of order l in $\cos\theta$ and $\sin\theta$, where a factor of $\sin\theta$ is accompanied by either $\cos\phi$ or $\sin\phi$. The terms with $l > 2$ are all very small, but it is of interest that modern satellite measurements have determined their strength up to approximately $l = 40$. Such terms will not be needed in our subsequent discussion, and we will use only the explicitly written terms in Eq. (2.9) from here on.

Centrifugal Effects The Earth is rotating, so a position fixed with respect to the Earth is in a rotating coordinate system and centrifugal effects must be taken into account. Recall that in elementary physics a mass μ, which is rotating at angular frequency ω experiences a "centrifugal force." For a mass located at a distance of $r_\perp$ from the axis of rotation, the centrifugal force is

$$\mu r_\perp \omega^2$$

and points away from the axis of rotation. The centrifugal force can be derived from a centrifugal potential, given by

$$V_{cent} = -\frac{1}{2} r_\perp^2 \omega^2. \tag{2.11}$$

Denoting the centrifugal force on a mass μ as $\boldsymbol{F}_{cent}$, we have

$$\boldsymbol{F}_{cent} = -\mu \boldsymbol{\nabla} V_{cent} = \mu \omega^2 \boldsymbol{r}_\perp. \tag{2.12}$$

Returning to the case of the Earth's rotation and using spherical coordinates, $r_\perp$ is just the distance from the Earth's axis and is given by $r_\perp = r \sin\theta$. The centrifugal potential becomes

$$V_{cent} = -\frac{1}{2}(r\omega \sin\theta)^2, \tag{2.13}$$

and $\omega = 2\pi/T_{sid}$. Adding V_{cent} to our expression for V_g from Eq. (2.9) and dropping all angular dependence from Legendre functions with $l > 2$, , the total potential is

$$V_{tot} = -\frac{GM_e}{r}[1 - \frac{1}{2} J_2 (\frac{a}{r})^2 (3\cos^2\theta - 1)] - \frac{1}{2}(r\omega \sin\theta)^2. \tag{2.14}$$

The local acceleration due to gravity and centrifugal effects is defined as

$$\boldsymbol{g} = -\boldsymbol{\nabla} V_{tot}. \tag{2.15}$$

Evaluating $\boldsymbol{g}$ using Eq. (2.15) at the equator, the spherically symmetric term in V_{tot} gives $\sim 980\,\mathrm{cm/s^2}$. The J_2 term is

$\sim 1.6\,\text{cm/s}^2$. The centrifugal term is $\sim 3.4\,\text{cm/s}^2$. We see that the spherically symmetric term is dominant, while the J_2 and centrifugal terms are much smaller.

It is important to note that in a rotating coordinate system, Newton's second law is modified in two ways, commonly known as centrifugal and Coriolis forces. The addition of V_{cent} to V_g does account for centrifugal effects, but it leaves out the Coriolis force. The latter comes into play only when the body being considered is moving relative to the Earth, e.g., a part of the atmosphere which is in motion as in a storm, discussed further in Sec. (6). For a body of mass μ hanging at rest from a string connected to an earthbound support, there is no Coriolis force and the tension in the string is given by

$$\boldsymbol{T} = -\mu \boldsymbol{\nabla} V_{tot} = \mu \boldsymbol{g}.$$

This provides a simple way to measure $\boldsymbol{g}$.

Equipotentials and Gravitational Acceleration What is commonly called Clairaut's theorem, is an equation for the gravitational acceleration on the Earth's surface as a function of latitude or equivalently, the polar angle θ. As will be seen below, it is easy to derive from our expression for V_{tot} as given in Eq. (2.14) and holds at the same level of accuracy. Using Eq. (2.15) to obtain $\boldsymbol{g}$ leads to an expression of the form

$$\boldsymbol{g} = g_r \hat{\boldsymbol{r}} + g_\theta \hat{\boldsymbol{\theta}}. \tag{2.16}$$

Both g_r and g_θ depend on r and θ.

To proceed, one needs to find the surface of the Earth. Specifically, for a given latitude or polar angle, what is the value of r that corresponds to being at the surface of the Earth? The answer to this question depends on the notion of equipotential, an equipotential being a surface on which

V_{tot} is constant. An important property of an equipotential is that the gravitational force on a mass is always perpendicular to the equipotential surface on which the mass is located. A fluid cannot support a force tangent to its surface, so the surface of a fluid which is in mechanical equilibrium will coincide with a gravitational equipotential. As a simple example of this, consider a handheld container of water. If the container is tilted, the surface of the water remains horizontal at a constant elevation h, and therefore at a constant value of the local gravitational potential, gh. Applying this same reasoning to the Earth as a whole, it is clear that the surface of the ocean must be an equipotential. Moving onto dry land, the average elevation over the surface of the Earth is $\sim 800\,\mathrm{m}$, so the value of V_{tot} increases on average by only one part in 10^4. This is a very small amount. It is then well within the accuracy of our formula for V_{tot} to assume that the Earth's surface is an equipotential.

Assuming this to be true, V_{tot} must take the same value at any point on its surface. In particular this includes the equator and the poles, where r takes the values a and b, respectively. We have

$$V_{tot}(a, 90^\circ) = V_{tot}(b, 0). \tag{2.17}$$

It will be useful to introduce a dimensionless parameter which characterizes the contribution of the centrifugal term. Following Fermi, we define m (not a mass) by

$$m = \frac{a^3\omega^2}{GM_e}. \tag{2.18}$$

We now have three small dimensionless parameters: f, J_2, and m. The formulas that follow will be correct to first order in these small parameters. Higher-order terms will be dropped. Writing out $V_{tot}(a, 90^\circ)$, we have

$$V_{tot}(a, 90^\circ) = -\frac{GM_e}{a}[1 + \frac{J_2}{2} + \frac{m}{2}]. \tag{2.19}$$

Doing the same at $r = b, \theta = 0$, we have

$$V_{tot}(b,0) = -\frac{GM_e}{b}[1 - J_2(\frac{a}{b})^2]. \tag{2.20}$$

We may simplify this expression by keeping only first-order terms. Since J_2 is already first order, we may make the replacement $a/b \to 1$ in its coefficient. Using the definition of the flattening from Eq. (2.1), we have $b/a = 1 - f$. Using this to rewrite the first term in Eq. (2.20), we have

$$\frac{GM_e}{b} = \frac{GM_e}{a}(\frac{a}{b}) = \frac{GM_e}{a}\frac{1}{1-f} \sim \frac{GM_e}{a}(1+f). \tag{2.21}$$

Our final, correct to first order, expression for $V_{tot}(b,0)$ is

$$V_{tot}(b,0) = -\frac{GM_e}{a}[1 + f - J_2]. \tag{2.22}$$

Equating this to $V_{tot}(a,0)$, we obtain

$$-\frac{GM_e}{a}[1 + f - J_2] = -\frac{GM_e}{a}[1 + \frac{J_2}{2} + \frac{m}{2}]. \tag{2.23}$$

Canceling terms, we finally have

$$f = \frac{3}{2}J_2 + \frac{1}{2}m, \tag{2.24}$$

so only two of our three small parameters are independent quantities. Using modern data, Eq. (2.24) is accurate to a part in a thousand, so the approximation of working to first order is a very good one.

Eq. (2.24) will be used in what follows, but we still must find the value of r that corresponds to being on the Earth's surface for an arbitrary polar angle θ. The work done above motivates the following trial form,

$$r(\theta) = a(1 - f\cos^2\theta). \tag{2.25}$$

This certainly works at $\theta = 0°, 90°$, and $180°$, but many other formulas would also work for these angles. What singles out Eq. (2.25) is that when $r(\theta)$ from Eq. (2.25) is substituted in Eq. (2.14) for V_{tot}, the result is constant, independent of θ, implying that an equipotential has been reached, to first order. The elementary steps showing this will not be written out, but it is worth noting that the cancellation of all factors of $\cos^2\theta$ is a direct consequence of Eq. (2.24).

Clairaut's Formula Knowing the parameters of the Earth's surface, we can now evaluate the acceleration due to gravity on it. What we want is the magnitude of the gravitational acceleration, as given by

$$g(\theta) \equiv \left[(\frac{\partial V_{tot}}{\partial r})^2 + (\frac{1}{r}\frac{\partial V_{tot}}{\partial \theta})^2 \right]^{1/2}, \qquad (2.26)$$

where r is evaluated using Eq. (2.25) after the derivatives are taken. Since the term involving $\partial/\partial\theta$ is the square of a first-order term and therefore of second order, the value of $g(\theta)$ correct to first order is given by the simpler formula

$$g(\theta) = |\frac{\partial V_{tot}}{\partial r}|. \qquad (2.27)$$

Evaluating the derivative, and substituting for $r(\theta)$ from Eq. (2.25), we obtain

$$g(\theta) = \frac{GM_e}{a^2}[(1 + \frac{3}{2}J_2 - m) + (2f - \frac{9}{2}J_2 + m)\cos^2\theta]. \qquad (2.28)$$

Eliminating J_2 using Eq. (2.24), we find

$$g(\theta) = \frac{GM_e}{a^2}[(1 + f - \frac{3}{2}m) + (\frac{5}{2}m - f)\cos^2\theta]. \qquad (2.29)$$

The equator is at $\theta = 90°$, so setting $g_{eq} = g(90°)$, we have

$$g_{eq} = \frac{GM_e}{a^2}(1 + f - \frac{3}{2}m), \qquad (2.30)$$

and, again dropping higher-order terms, $g(\theta)$ becomes

$$g(\theta) = g_{eq}[1 + (\frac{5}{2}m - f)\cos^2\theta]. \tag{2.31}$$

This is Clairaut's formula. As treated here, it is a first-order expression so not really a "theorem." Pursuing this point a bit further, Clairaut's formula can be an exact result or theorem if certain detailed assumptions are made about the Earth's internal structure. It is in the spirit of Fermi's approach to physics to derive it as we have done from the form of the gravitational potential near the Earth, rather than to model the interior of the Earth. Nevertheless, theorem or formula, it is a remarkable expression. As we will see in the next section, it gives a quite accurate description of the acceleration due to gravity at various points on the Earth.

The Accuracy of Clairaut's Formula Clairaut's formula is remarkably accurate for the variation of g over the Earth. Using modern values of g_e, m and f taken from Table 2 of (Yoder 1995; Ahrens 1995), and evaluating $g(\theta)$ using Eq. (2.27), Clairaut's formula for $g(\theta)$ becomes

$$g(\theta) = 978.03 + 5.18\cos^2\theta, \tag{2.32}$$

where $g(\theta)$ is expressed in cm/s^2, and we have retained five significant figures. Recalling that the latitude is $|\theta - 90°|$, Eq. (2.32) is used to compute the local value of $g(\theta)$ for several locations.

Table 2.2: Accuracy of Clairaut's theorem for several locations

Location	Latitude	Measured	Clairaut
North Pole	90° N	983.2	983.21
Stockholm	59° N	981.8	981.84
New York	41° N	980.3	980.26
Auckland	37° S	980.0	979.91

From Table (2.2) it is seen that Clairaut's theorem is accurate at the 0.1% level.

2.3 Pressure at the Center of the Earth

Fermi also does a simple calculation of the pressure at the center of the Earth. The Earth's rotation and slight non-sphericity are very small effects here and are ignored. The Earth is taken to be spherically symmetric, of radius R_e, and in hydrostatic equilibrium. The equation of hydrostatic equilibrium for a point inside the Earth is

$$-\boldsymbol{\nabla} p + \rho \boldsymbol{g} = 0, \tag{2.33}$$

where p is the pressure, $\boldsymbol{g}$ is the acceleration due to gravity, and ρ is the mass density. The acceleration is related to the gravitational potential by

$$\boldsymbol{g} = -\boldsymbol{\nabla} V_g. \tag{2.34}$$

For a spherically symmetric Earth, we have

$$\boldsymbol{g} = -\hat{\boldsymbol{r}} g(r), \tag{2.35}$$

where

$$g(r) = \frac{GM(r)}{r^2}, \tag{2.36}$$

and $M(r)$ is the mass inside a sphere of radius r,

$$M(r) = 4\pi \int_0^r \rho(r) r^2 dr. \tag{2.37}$$

Evaluation of this integral requires a model for $\rho(r)$. The simplest possible assumption would be to use the average density of the Earth for $\rho(r)$. However, this ignores the fact that the density at the center of the Earth must be significantly larger than the density at the surface. Fermi chooses the next simplest model by taking

$$\rho(r) = \rho_0 - \rho_1 r. \tag{2.38}$$

(We slightly changed Fermi's notation, making the replacement $p \to \rho_1$ in his formula for $\rho(r)$.) To determine ρ_0 and ρ_1, Fermi uses the average mass density of $5.5\,\mathrm{g/cm^3}$, and a surface mass density of $\rho(R_e) = 3\,\mathrm{g/cm^3}$. This results in $\rho_0 = 13\,\mathrm{g/cm^3}$, and $\rho_1 R_e = 10\,\mathrm{g/cm^3}$.

The equation of hydrostatic equilibrium is now

$$\frac{\partial p}{\partial r} = -\frac{4\pi G \rho(r)}{r^2} \int_0^r \rho(r') r'^2 dr'.$$

Doing the integral, we have

$$\frac{\partial p}{\partial r} = -\frac{4\pi G}{r^2} (\rho_0 - \rho_1 r)(\frac{1}{3}\rho_0 r^3 - \frac{1}{4}\rho_1 r^4).$$

Integrating again from $r = 0$ to R_e, we arrive at Fermi's formula

$$p(0) - p(R_e) = \frac{2}{3}\pi G(\rho_0 R_e)^2 \left(1 - \frac{7}{6}(\frac{\rho_1 R_e}{\rho_0}) + \frac{3}{8}(\frac{\rho_1 R_e}{\rho_0})^2\right).$$

The (atmospheric) pressure at the Earth's surface can be ignored compared to the pressure at the center. For numerical evaluation, we use the SI unit of pascals (Pa) where $1\,\text{Pa} = 1\,\text{N/m}^2$. Fermi's formula gives

$$p(0) = 3.11 \times 10^8\,\text{kPa} = 3.07 \times 10^6\,\text{atm},$$

where we use 1 atm = 101,325 Pa. The amount of information on the interior of the Earth which is available at present compared to Fermi's time is truly enormous. Nevertheless, Fermi's simple approach yields a number fairly close to $3.6 \times 10^8\,\text{kPa}$, the current estimate of the pressure at the center of the Earth.

2.4 The Moments of Inertia of the Earth

Continuing, Fermi returns to the gravitational potential and its connection with the moments of inertia of the Earth. In order to make contact with modern results on the gravitational potential, we briefly give a more general discussion before returning to Fermi's notes. We start from a general expression for the gravitational potential,

$$V_g(\boldsymbol{r}) = -G \int d^3r' \frac{\rho(\boldsymbol{r}')}{|\boldsymbol{r} - \boldsymbol{r}'|}. \tag{2.39}$$

This is just Newton's gravitational potential for a continuous mass distribution, $\rho(\boldsymbol{r})$. At any point outside of the Earth, $|\boldsymbol{r}| > |\boldsymbol{r}'|$, and it is useful to expand the gravitational potential in inverse powers of $r = |\boldsymbol{r}|$. Starting with the expansion of the denominator in Eq. (2.39), we have

$$\frac{1}{|\boldsymbol{r} - \boldsymbol{r}'|} = \frac{1}{r}\left[1 - 2\frac{\boldsymbol{r}\cdot\boldsymbol{r}'}{r^2} + (\frac{r'}{r})^2\right]^{-1/2} \tag{2.40}$$

$$= \frac{1}{r}\left[1 + \frac{\boldsymbol{r}\cdot\boldsymbol{r}'}{r^2} + \frac{1}{2r^4}[3(\boldsymbol{r}\cdot\boldsymbol{r}')^2 - r^2(r')^2] + \ldots\right].$$

Inserting this expansion into Eq. (2.39), we may write V_g as a series,

$$V_g(\boldsymbol{r}) = V_g^{(0)}(\boldsymbol{r}) + V_g^{(1)}(\boldsymbol{r}) + V_g^{(2)}(\boldsymbol{r}) + \ldots, \tag{2.41}$$

where

$$V_g^{(0)}(\boldsymbol{r}) = -\frac{G}{r}\int d^3r'\rho(\boldsymbol{r}') = -\frac{GM_e}{r}, \tag{2.42}$$

$$V_g^{(1)}(\boldsymbol{r}) = -(\frac{G}{r^3})r_j \int d^3r'\rho(\boldsymbol{r}')r_j', \tag{2.43}$$

$$V_g^{(2)}(\boldsymbol{r}) = -(\frac{G}{2r^5})(3r_jr_k - r^2\delta_{jk})\int d^3r'\rho(\boldsymbol{r}')r_j'r_k', \text{ etc.,} \tag{2.44}$$

with repeated indices being summed in all cases. In general, $V_g^{(l)}(\boldsymbol{r})$ varies with r as $1/r^{l+1}$. Choosing the origin at the Earth's center of mass, the $l = 1$ term vanishes, so the first nonvanishing term has $l = 2$. It can be expressed in terms of the Earth's moment of inertia.

The moment of inertia tensor is defined as

$$I_{jk} = \int \rho(\boldsymbol{r}')(\delta_{jk}(r')^2 - r_j'r_k')d^3r. \tag{2.45}$$

Comparing this integral to the one in $V_g^{(2)}(\boldsymbol{r})$, the $\delta_{jk}(r')^2$ term in Eq. (2.45) is absent in the $V_g^{(2)}(\boldsymbol{r})$ integral. However, since

$$(3r_jr_k - r^2\delta_{jk})\delta_{jk} = 0,$$

the same answer is obtained if we replace $r_j'r_k'$ with $r_j'r_k' - (r')^2\delta_{jk}$ in the integrand of $V_g^{(2)}(\boldsymbol{r})$. After doing that, we have

$$V_g^{(2)}(\boldsymbol{r}) = (\frac{G}{2r^5})(3r_jr_k - r^2\delta_{jk})I_{jk}. \tag{2.46}$$

This formula makes it clear that $V_g^{(2)}$ is determined by the moment of inertia tensor. Although we will not consider l values greater than 2, each term in $V_g^{(l)}$ is determined by a corresponding moment, involving an integral of the density ρ weighted with l powers of r.

The situation considered by Fermi assumes the Earth is azimuthally symmetric. For an azimuthally symmetric body, the moment of inertia tensor is diagonal. The polar axis element I_{33} is conventionally denoted as C. The elements I_{11} and I_{22} are equal, and their common value is conventionally denoted as A. The $l = 2$ term becomes

$$V_g^{(2)} = \frac{G(C-A)}{2r^5}(3r_3^2 - r^2) = \frac{G(C-A)}{2r^3}(3\cos^2\theta - 1), \tag{2.47}$$

where the angle θ is measured with respect to the Earth's polar axis. Writing out V_g, including $l = 0$ and $l = 2$ terms, we have

$$V_g = -\frac{GM_e}{r} + \frac{G(C-A)}{2r^3}(3\cos^2\theta - 1). \tag{2.48}$$

The form of $V_g^{(2)}$ given in Eqs. (2.47) and (2.48) is known as McCullagh's formula. Comparing to Eq. (2.9), we see that

$$J_2 = \frac{C-A}{M_e a^2}. \tag{2.49}$$

The values of C and A are well known (Yoder 1995) and can be written as multiples of $M_e a^2$, where M_e is the Earth's mass and a is the equatorial radius. Using that notation, the values of A and C are

$$A = 0.3296108\, M_e a^2, \quad C = 0.3307007\, M_e a^2. \tag{2.50}$$

For comparison, if the Earth were an axially symmetric ellipsoid with a uniform mass density, we would obtain

$$A_{el} = 0.4000\, M_e a^2, \quad C_{el} = 0.3986\, M_e a^2. \tag{2.51}$$

We see that both of these values are larger than the actual values of A and C. This actual data is consistent with the Earth having a nonuniform mass density, with higher density closer to the center of the Earth. Using the known values of A and C to compute J_2 via Eq. (2.49) gives $J_2 = 1.089 \times 10^{-3}$, which is within 1% of the value determined by satellite observations of the Earth's gravitational potential.

The small inhomogeneities of the Earth's mass distribution show up in the higher l terms in the gravitational potential. Modern measurements of the gravitational potential make use of very accurate satellite data and are able to determine terms up to $l \sim 40$ in the gravitational potential (Nerem 1994).

2.5 Motions of the Earth's Axis

In the next two sections, we treat material not explicitly covered in Fermi's lecture notes: the precession of the equinoxes and the search for Euler's nutation. This material is of long-term historical interest and fits naturally onto subjects just covered in preceding sections.

2.6 Equinoxes and Solstices

The existence of seasons certainly affects the physics of the Earth's surface. The cause of seasons is that the Earth's spin axis is tilted at an angle with respect to the normal to the ecliptic plane, the plane of the Earth's orbit. The tilt angle is $\theta_t = 24.44°$. The onsets of spring and fall occur at the vernal and autumnal equinoxes, respectively. As the names suggest, night and day have the same duration at an equinox. Between the autumnal and vernal equinoxes there is the winter solstice, when the the Earth's axis points the maximal amount away from the Sun. Similarly, the summer solstice occurs between the vernal and autumnal equinoxes, when the the Earth's axis points the maximal amount toward the Sun. The winter and summer solstices are often referred to as the "shortest" and "longest" days of the year, the reference being to the number of hours of sunlight in the Northern Hemisphere.

To define the locations of equinoxes and solstices precisely, let $\hat{\boldsymbol{n}}$ be a unit vector along the Earth's spin axis, and let $\hat{\boldsymbol{n}}_{ec}$ be the projection of that vector on the ecliptic plane. At a solstice $\hat{\boldsymbol{n}}_{ec}$ is parallel or antiparallel to $\boldsymbol{R}_{se}$, the vector from the center of mass of the Sun to the center of mass of the Earth. An equinox occurs when $\hat{\boldsymbol{n}}_{ec}$ is perpendicular to $\boldsymbol{R}_{se}$. The Earth's position at the various seasons is shown schematically in Fig. (2.1). The vertical arrows are along $\hat{\boldsymbol{e}}$, the normal to the ecliptic plane. The slanted arrows are pointed along the Earth's spin axis, $\hat{\boldsymbol{n}}$. The $\hat{\boldsymbol{n}}$ vector rotates very slowly around $\hat{\boldsymbol{e}}$, the period of that rotation being many thousands of years. Determining the period of that rotation is the subject of the next section.

2.7 Precession of the Equinoxes

As discussed in previous sections, the Earth is slightly nonspherical. This means that in addition to attracting the Earth, the Sun and Moon exert torques on the Earth. The net torque causes the Earth's angular momentum vector to precess very slowly about the normal to the ecliptic plane. The resulting slow motion along the Earth's orbit is called "precession of the equinoxes." Of course the solstices change location along the Earth's orbit as well, but "precession of the equinoxes" has been used for this precession. To visualize the effect of this precession, let T_{ex} be its period. After passage

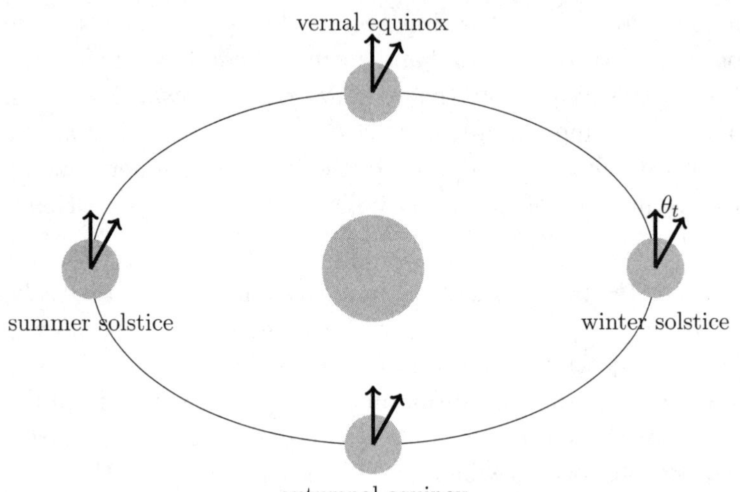

Figure 2.1: The Earth's spin axis at various seasons

of times $T_{ex}/4$ or $3T_{ex}/4$, equinoxes and solstices will be interchanged, so March 21, instead of being the vernal equinox, will be the winter or summer solstice. Likewise, after a time of $T_{ex}/2$, equinoxes remain equinoxes, but winter and summer solstices are interchanged. History recorded so far has not been long enough to see these dramatic effects, but the small shifts in the dates of equinoxes and solstices have been known from the time of the ancient Greeks.

To compute the precession rate, we turn to the angular equation of motion

$$\boldsymbol{K} = \frac{d\boldsymbol{L}}{dt}, \tag{2.52}$$

where the angular momentum of the Earth $\boldsymbol{L}$, like the torque $\boldsymbol{K}$, is computed about the Earth's center of mass. The angular momentum of the Earth is completely dominated by the spinning of the Earth on its axis, since the shortest period in the problem by far is the length of a single day, T_{sol}. Defining

$$\omega_d \equiv \frac{2\pi}{T_{sol}}, \tag{2.53}$$

with negligible error we have

$$\boldsymbol{L} = (C\omega_d)\hat{\boldsymbol{n}}, \tag{2.54}$$

where C is the moment of inertia about the Earth's spin axis. It is useful to break $\boldsymbol{L}$ into two components, one along the normal to the ecliptic plane, $\boldsymbol{L}_{\parallel}$, and the other perpendicular to the normal, and hence in the ecliptic plane, $\boldsymbol{L}_{\perp}$. Their magnitudes are given by

$$|\boldsymbol{L}_{\parallel}| = C\omega_d \cos\theta_t, \quad |\boldsymbol{L}_{\perp}| = C\omega_d \sin\theta_t. \tag{2.55}$$

The solution of Eq. (2.52) becomes very simple if we average both sides over a period long compared to a year or a month, but much shorter than the many thousands of years involved in the period of the equinox precession. The time-averaged torque $\bar{\boldsymbol{K}}$ due to the Sun and the Moon is calculated in Appendix B, Eq. (B.13). As shown there, the magnitude of $\bar{\boldsymbol{K}}$ is

$$|\bar{\boldsymbol{K}}| = \frac{3}{2}\sin\theta_t \cos\theta_t (C - A)(\frac{GM_s}{R_{se}^3} + \frac{GM_m}{R_{me}^3}), \tag{2.56}$$

and it lies in the ecliptic plane, perpendicular to $\boldsymbol{L}_{\perp}$. The torque has no effect on $\boldsymbol{L}_{\parallel}$, which is therefore constant. The torque rotates $\boldsymbol{L}_{\perp}$, leaving the magnitude of $\boldsymbol{L}_{\perp}$ unchanged. Let the angular frequency with which $\boldsymbol{L}_{\perp}$ rotates be ω_{ex}. This is the sought-for frequency of the equinox precession. We then have

$$|\frac{d}{dt}\boldsymbol{L}_{\perp}| = \omega_{ex}|\boldsymbol{L}_{\perp}| = \omega_{ex}(C\omega_d \sin\theta_t). \tag{2.57}$$

The angular equation for $|\boldsymbol{L}_\perp|$ is

$$|\bar{\boldsymbol{K}}| = |\frac{d}{dt}\boldsymbol{L}_\perp|. \tag{2.58}$$

Writing this out using Eqs. (2.56) and(2.57), we have

$$\frac{3}{2}\sin\theta_t\cos\theta_t(C - A)(\frac{GM_s}{R_{se}^3} + \frac{GM_m}{R_{me}^3}) = \omega_{ex}(C\omega_d\sin\theta_t). \tag{2.59}$$

Solving for the equinox precession frequency, we find

$$\omega_{ex} = \frac{3}{2}(\frac{C - A}{C})\frac{\cos\theta_t}{\omega_d}(\frac{GM_s}{R_{se}^3} + \frac{GM_m}{R_{me}^3}) \tag{2.60}$$

Using current values for the various parameters in Eq. (2.60), we have

$$\omega_{ex} = 8.26 \times 10^{-12}\,\text{rad/s}, \tag{2.61}$$

corresponding to a period of 24,200 years, within a few percent of the actual value of 25,800 years (Yoder 1995). The discrepancy is largely accounted for by taking account of the eccentricities of the Earth and Moon's orbits. It is worth pointing out that the sense of the precession about the normal to the ecliptic plane is opposite to the sense of the Earth's orbit around the Sun. Viewing the ecliptic plane from above, the Earth's orbit around the Sun has a counterclockwise sense, while the precession of the Earth's axis about the ecliptic plane normal is clockwise. The effect is to make each season appear a tiny bit sooner each year. Of course, once the period of the precession of the equinoxes has elapsed, the Earth's axis has turned through a full 360° and returns to its original position. The precession of the equinoxes was observed over 2000 years ago.

The first quantitative description is usually attributed to Hipparchus in the first century BCE but its explanation had to wait until Newton's great work on motion and gravity in 1687.

2.8 Nutation and Torque-Free Rotational Motion

In classical mechanics (Goldstein 1999; Landau and Lifshitz 1976), the term "nutation" refers to an oscillation in what is usually known as the "second Euler angle." This was θ_t in Fig. (2.1), the angle between the Earth's symmetry axis and the normal to the ecliptic plane. As the Earth undergoes the precession of the equinoxes it also undergoes oscillation or nutation in θ_t. One of the main contributors to nutation of the Earth is the variation with time of the torque on the Earth due to the Moon. This has a period of 18.6 years.

In astronomy, the term "nutation" is sometimes used in a broader sense. An interesting and easy to understand example is a phenomenon which has various names, "Euler's nutation," "free nutation," and "polar motion." Up to now, we have assumed that the Earth's axis of symmetry is coincident with the direction of the Earth's angular momentum $\boldsymbol{L}$. However, observations involving the fixed stars show that there is a tiny angle between the Earth's symmetry axis and its angular momentum vector, 9.7×10^{-7} rad. If two lines are drawn through the center of mass of the Earth, one along the symmetry axis and the other parallel to $\boldsymbol{L}$, they cross the Earth's surface near the North Pole at points only $\sim 6\,\text{m}$ apart, certainly a small distance on an astronomical scale!

In this section, we will ignore external torques and treat the angular momentum of the Earth as a constant vector. This may seem paradoxical given that in the preceding section, the precession of the equinoxes was explained as the direct result of external torques, but this is not as strange as it seems at first glance. The torques which act on the Earth from other planets or the Sun are very small, and to be observable, time intervals of thousands of years are needed. In comparison, the time intervals of interest in Euler's nutation are very short (300–400 days). Over such small time intervals, the torques due to other planets only change the angular momentum by a negligible amount and therefore it is an excellent approximation to treat the Earth's angular momentum $\boldsymbol{L}$ as constant over intervals of 300–400 days.

We will make use of two coordinate systems: one which is space-fixed (xyz) with $\boldsymbol{L}$ along the z axis and one which is body-fixed (123) with the 3 axis along the Earth's axis of symmetry. The space-fixed system is generated from the one used in Sec. (2.7) by a rotation which takes the normal to the ecliptic plane to a direction parallel to $\boldsymbol{L}$. The body-fixed coordinate system is a rotating coordinate system, rotating at angular velocity $\boldsymbol{\omega}$, which in components is

$$\boldsymbol{\omega} = \omega_1 \hat{\mathbf{1}} + \omega_2 \hat{\mathbf{2}} + \omega_3 \hat{\mathbf{3}}. \tag{2.62}$$

As explained in more detail in Sec. (6), in computing the time rate of change of any vector, it is necessary to distinguish between the rate of change in the space-fixed coordinate system and that in the body-fixed system. The relation between the two is[2]

$$(\frac{d\boldsymbol{L}}{dt})_{sf} = (\frac{d\boldsymbol{L}}{dt})_{bf} + \boldsymbol{\omega} \wedge \boldsymbol{L}. \tag{2.63}$$

In the absence of torques, the angular momentum in the space-fixed system is conserved, so Eq. (2.63) is now

$$0 = (\frac{d\boldsymbol{L}}{dt})_{bf} + \boldsymbol{\omega} \wedge \boldsymbol{L}. \tag{2.64}$$

Written out in components, Eq. (2.64) becomes a set of three equations known as Euler's equations. For the case of the Earth with $I_3 = C$ and $I_1 = I_2 = A$, the angular momentum is

$$\boldsymbol{L} = A\omega_1 \hat{\mathbf{1}} + A\omega_2 \hat{\mathbf{2}} + C\omega_3 \hat{\mathbf{3}}, \tag{2.65}$$

[2]For vectors $\boldsymbol{A}$ and $\boldsymbol{B}$, $\boldsymbol{A} \wedge \boldsymbol{B}$ is the cross-product, equal to $\boldsymbol{A} \times \boldsymbol{B}$.

and Euler's equations are

$$
\begin{aligned}
0 &= A\dot{\omega}_1 + \omega_2\omega_3(C - A) \\
0 &= A\dot{\omega}_2 - \omega_1\omega_3(C - A) \\
0 &= C\dot{\omega}_3.
\end{aligned}
\tag{2.66}
$$

From the third of Eqs. (2.66), we learn that ω_3 is constant and therefore so is $L_3 = C\omega_3$. Defining β as the angle between $\boldsymbol{L}$ and the 3 axis, we have

$$
L_3 = \boldsymbol{L} \cdot \hat{\mathbf{3}} = |\boldsymbol{L}| \cos \beta. \tag{2.67}
$$

It follows that the angle between the angular momentum vector and the Earth's axis of symmetry is constant. Fig. (2.2) shows the angle β and indicates the sense of the Earth's rotation around its axis of symmetry. The setup is similar to that shown in Fig. (2.1), except that here the z axis is along $\boldsymbol{L}$, rather than along the normal to the ecliptic plane.

Defining the part of $\boldsymbol{L}$ which is perpendicular to the symmetry axis as $\boldsymbol{L}_\perp = A(\omega_1 \hat{\mathbf{1}} + \omega_2 \hat{\mathbf{2}})$, the angular momentum vector is $\boldsymbol{L} = L_3 \hat{\mathbf{3}} + \boldsymbol{L}_\perp$, and its square is

$$
\boldsymbol{L} \cdot \boldsymbol{L} = (L_3)^2 + \boldsymbol{L}_\perp \cdot \boldsymbol{L}_\perp. \tag{2.68}
$$

Eq. (2.68) shows that

$$
\boldsymbol{L}_\perp \cdot \boldsymbol{L}_\perp = A^2(\omega_1^2 + \omega_2^2)
$$

is constant, so $\omega_1^2 + \omega_2^2$ is also constant. To the Earth-fixed observer, the angular momentum appears to whirl around the symmetry axis, L_3 and $|\boldsymbol{L}_\perp|$ remaining constant, with the direction of $\boldsymbol{L}_\perp$ moving in a circle. This is Euler's nutation. Defining the corresponding frequency as

$$
\omega_{eu} \equiv \omega_3(\frac{C - A}{A}), \tag{2.69}
$$

the first two Euler equations become

$$\dot{\omega}_1 = -\omega_{eu}\,\omega_2 \qquad (2.70)$$
$$\dot{\omega}_2 = +\omega_{eu}\,\omega_1.$$

To an Earth-fixed observer, Eqs. (2.70) describe a counter-clockwise motion of the angular momentum around the Earth's symmetry axis. (Note that for a prolate object with $C < A$, the motion would be clockwise.)

To evaluate ω_{eu}, the modern values of C and A were given in Eq. (2.50). Using these we have

$$\frac{C-A}{A} = 0.0033066 = \frac{1}{302.423}. \qquad (2.71)$$

The frequency ω_3 is related to the length of a day by $\omega_3 = 2\pi/T_d$, and likewise ω_{eu} is related to its period by $\omega_{eu} = 2\pi/T_{eu}$. Rewriting Eq. (2.69), we have

$$\frac{2\pi}{T_{eu}} = \frac{2\pi}{T_d}(\frac{C-A}{A}), \text{ or } T_{eu} = T_d(\frac{A}{C-A}). \qquad (2.72)$$

Since T_d is one day, using Eq. (2.71), we find a period for Euler's nutation of

$$T_{euler} = 302.423\,\text{days}. \qquad (2.73)$$

An intensive search for a nutation with this period was carried out in the late nineteenth century. The results were negative. In 1891 Seth C. Chandler, an American amateur astronomer, investigated a wider range of periods and discovered a nutation with a period of approximately 434 days. This has come to be known as the "Chandler wobble." A prime suspect for the reason the period is so different from the value calculated by Euler is that the assumption that the Earth is a rigid body has finally broken down. This is plausible, given the presence of the oceans near the surface of the Earth, and regions of the Earth's interior which are liquid. However, this explanation encounters its own problems, namely, if the period of the Chandler wobble is largely due to the fluidity, the wobble itself should die away in less than 100 years. It remains a fascinating problem still of interest today to explain how a torque-free Earth and accompanying atmosphere can maintain an enduring wobble with the correct frequency.

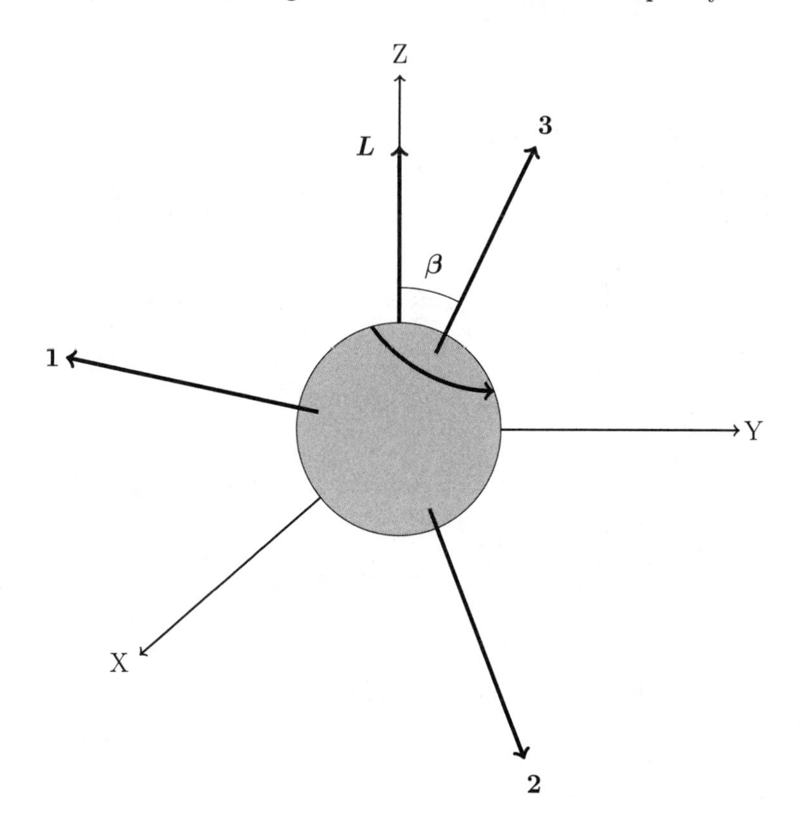

Figure 2.2: Axis and angular momentum of the Earth

Equilibrium of the atmosphere (3)

$$\frac{dp}{dz} = -\rho g \qquad pV = RT$$

$$\rho = \frac{M}{V} = \frac{M}{RT} p$$

$$\frac{dp}{dz} = -\frac{\rho}{RT} p$$

Isothermal atmosphere

$$p = p_0 e^{-\frac{gMz}{RT}}.$$

Distribution of gases at different heights

$M = 28.88$

$R = 8.2 \times 10^7$

$T = 273$

$$\frac{RT}{Mg} = 7.8 \times 10^5 \text{ cm} = 7.8 \text{ Km}$$

Adiabatic atmosphere

$\kappa = 1.4$

$$pV^\kappa = RT \text{ const} \qquad p \sim \rho^\kappa$$

$$T^\kappa / p^{\kappa-1} = \text{const} \qquad \rho = \rho_0 \frac{1}{p_0} p^{1/\kappa}$$

$$T \sim p^{\frac{\kappa-1}{\kappa}}$$

$$\frac{dp}{dz} = -\frac{g\rho_0}{p_0^{1/\kappa}} p^{1/\kappa}$$

H. ~~Heights of adiabatic atmosphere~~

$$\int \frac{dp}{p^{1/\kappa}} = \frac{\kappa}{\kappa-1}\left(p^{\frac{\kappa-1}{\kappa}} - p_0^{\frac{\kappa-1}{\kappa}}\right) = -\frac{g\rho_0}{p_0^{1/\kappa}} z$$

$$p^{\frac{\kappa-1}{\kappa}} = p_0^{\frac{\kappa-1}{\kappa}} - \frac{\kappa-1}{\kappa} \frac{g\rho_0}{p_0^{1/\kappa}} z$$

$$T = T_0\left(1 - \frac{\kappa-1}{\kappa} \frac{g\rho_0 z}{p_0}\right)$$

Atmospheric pressure vs. altitude

CHAPTER 3

Thermodynamics of the Earth's Atmosphere

In this chapter, we intersperse discussion of the contents of Fermi's notes with sections which review key topics in thermodynamics. Secs. (3.1)–(3.3), (3.5), and (3.6) review thermodynamics, while Secs. (3.4) and (3.7)–(3.10) expand on the actual material in Fermi's notes. More details on the subject of thermodynamics can be found in Fermi's own volume on the subject (Fermi 1956).

Thermodynamics, a field whose foundations were laid in the nineteenth century, is founded on the realization that heat is a form of energy and on the transformation of mechanical work into heat and of heat into mechanical work. It was further elucidated by the development toward the end of the nineteenth century of statistical mechanics. Statistical mechanics demonstrated how the heat content of a gas could be understood as the energy of motion of the molecules in the gas.

3.1 First Law of Thermodynamics

The first law of thermodynamics is a statement of energy conservation as applied to thermal systems. For the most part, in Fermi's notes, the system is the atmosphere, a gas. The internal energy of a gas or liquid is simply the total kinetic and potential energy of the system of molecules present. The internal energy is denoted as U. The system may expand (do work) or be compressed (be worked on). Likewise, the system may absorb or give off heat. If ΔQ is the heat absorbed in a thermodynamic process and ΔW is the work done by the system in the same process, then conservation of energy states

$$\Delta U = \Delta Q - \Delta W. \tag{3.1}$$

Eq. (3.1) is quite general, requiring only that each term in the equation is well defined. The process may involve equilibrium or nonequilibrium states, and the equation is independent of the particular variables needed to specify internal energy, heat, and work. In the next section, for processes connecting thermodynamic equilibrium states, it is shown how to write the equation in terms of differential changes in thermodynamic variables.

3.2 State Variables. Entropy

The equilibrium state of a thermodynamic system is defined by the values of a small number of macroscopic variables. For a gas containing a single type of molecule, the relevant variables are pressure, volume, temperature, and the number of molecules in the system. Any three of these are enough to specify the thermodynamic state of the system, the fourth being specified by the equation of state. State variables are independent of the history of the system and are either extensive or intensive. Extensive variables, like volume and number of particles, are "how much" variables. Intensive variables like pressure and temperature, are "how strong" variables.

Neither work nor heat is a state variable. Both are physical quantities whose values depend on the history of the system. Taking a system through a closed cycle of thermodynamic states, a net amount of work may be done, and heat absorbed or given off. But the system could equally well have been left in its initial state, and not taken through a cycle. In that case, the net work done and heat absorbed are zero, so the work done and the heat absorbed obviously depend on the history of the system, not just its present state.

The work done by a gaseous system in going from state 1 to state 2 is given by

$$W_{1\to 2} = \int_1^2 PdV. \tag{3.2}$$

The value of $W_{1\to 2}$ depends on the particular path taken through thermodynamic states in going from state 1 to state 2. For an infinitesimal change, $dW = PdV$.

There is a state variable S, known as entropy, whose integral with respect to temperature expresses the heat absorbed by the system. The concept of entropy was developed in the nineteenth century by three of the founders of the study of thermodynamics; Sadi Carnot, Rudolf Clausius, and William Thomson (Lord Kelvin), who worked in France, Germany, and Britain, respectively. Stated as an equation, the heat absorbed in a process which starts in state 1 and ends in state

2 is[1]

$$Q_{1\to 2} = \int_1^2 TdS. \tag{3.3}$$

The heat absorbed in a transition between two thermodynamic states depends on the path taken from state 1 to state 2, showing that $Q_{1\to 2}$ is also not a state variable. For an infinitesimal change,

$$dQ = TdS. \tag{3.4}$$

Entropy, like volume, number of particles, and internal energy, is an extensive variable. The concept of entropy has wide applicability beyond the realm of equilibrium thermodynamics discussed here. It is commonly regarded as a measure of the disorder in a system; the higher the entropy, the more disordered the system. For a thermodynamic system in equilibrium, it is generally true that the entropy increases when the temperature of the system increases.

We may now recast the first law in differential form for infinitesimal processes that do not involve molecules either entering or leaving the system. This is

$$dU = TdS - PdV. \tag{3.5}$$

In Sec. (3.5), we will generalize Eq. (3.5) to include the flow of molecules in or out of the system.

3.3 The Second Law of Thermodynamics. Carnot Cycle

A formulation of the second law, due to Lord Kelvin, says that heat extracted from a source kept at constant temper-

[1]The temperature T in Eq. (3.3) is the absolute temperature; such that absolute zero, approximately $-273°$C, is the lowest obtainable temperature.

ature cannot be fully transformed into work. Another one, first put forward by Clausius, says that it is impossible to have a transformation whose only function is to transfer heat from a body at lower temperature to one at higher temperature. The ideas associated with the second law are perhaps best illustrated by making use of a process first considered by Carnot, and known ever since as the Carnot cycle. Although not strictly needed to follow Fermi's lecture notes, a discussion of the Carnot cycle will allow us in Sec. (6.3) to discuss a modern subject that surely would have been of interest to Fermi: the use of Carnot cycles to understand the physics of hurricanes.

Carnot Cycle A Carnot cycle is a reversible cycle that takes a gas through a sequence of four steps. At the end the gas has returned to its original state. The cycle operates between two heat baths, one at absolute temperature $T_>$, and the other at absolute temperature $T_<$, with $T_>$ greater than $T_<$. In the first step of the cycle, the gas is placed in contact with the heat bath at $T_>$ and allowed to expand. In doing so, the gas performs work and absorbs an amount of heat $Q_>$, with the work performed equal to $Q_>$. Since it is at constant temperature, the internal energy of the gas remains constant. In the second step, the gas is removed from the heat bath and allowed to expand a second time, this time adiabatically, so no heat is transferred to or from the gas. Again, in expanding, the gas performs work. In this step, since there is no heat transfer, the gas lowers its internal energy, and hence its temperature, arriving at the lower value, $T_<$. In the third step, the gas is placed in contact with a heat bath at $T_<$ and compressed. Since work is performed on the gas in this step, it gives off heat $Q_<$, where $Q_<$ is equal to the work performed. Finally, in the last step, the gas is again isolated and adiabatically compressed back to its original state at the higher temperature, $T_>$.

In the first two steps of the cycle, the gas does work, and in the last two steps of the cycle, work is done on the gas. The difference between the work the gas does and the work performed on it is the net work, W. By conservation of energy, the net work is the heat absorbed minus the heat given off,

$$W = Q_> - Q_<. \tag{3.6}$$

The efficiency of the cycle is the ratio of the net work to the heat absorbed,

$$\eta = \frac{W}{Q_>} = 1 - \frac{Q_<}{Q_>}. \tag{3.7}$$

The efficiency can be expressed solely in terms of $T_>$ and $T_<$ using the fact that the entropy of the system is a state variable. In the isothermal expansion at temperature $T_>$, the entropy of the system increases by $Q_>/T_>$. The entropy does not change in the following adiabatic expansion. Then in the isothermal compression at temperature $T_<$, the entropy decreases by $Q_</T_<$. In the final adiabatic compression the entropy does not change. Since the entropy is a state variable, it must return to its initial value at the end of the cycle, so the gain in the isothermal expansion at $T_>$ must be balanced by the decrease in the isothermal expansion at $T_<$, so

$$\frac{Q_>}{T_>} = \frac{Q_<}{T_<}, \tag{3.8}$$

or

$$\frac{Q_<}{Q_>} = \frac{T_<}{T_>}. \tag{3.9}$$

Using this in Eq. (3.7), we finally have

$$\eta = 1 - \frac{T_<}{T_>}. \tag{3.10}$$

From Eq. (3.10), it is seen that a unit efficiency is only possible if $T_< = 0$, absolute zero.

3.4 Ideal Gas. Isothermal and Adiabatic Atmospheres

A substance in a gaseous state has the characteristic property that it fills the volume of its container regardless of the shape of the container. In our present discussion the gas in question is the Earth's atmosphere, a gas composed approximately of 78% diatomic nitrogen (N_2), 21% diatomic oxygen (O_2), and 0.93% argon (Ar), along with trace amounts of neon (Ne), helium (He), methane (CH_4), water vapor (H_2O), krypton (Kr), and diatomic hydrogen (H_2). We begin by treating "dry air," "moist air" referring to the case when a significant fraction of water molecules are present. The effects of moisture will be discussed in Sec. (3.7).

The equation of state of a gas is the relation that holds between its pressure, density, and temperature. An ideal gas is one in which thc interaction energy between molecules can be ignored compared to the molecular kinetic energy. Any gas becomes ideal at high enough temperature and low enough density. The Earth's atmosphere meets these conditions sufficiently well that treating it as an ideal gas is a good approximation. The equation of state of an ideal gas is

$$PV = N_{mol}RT. \tag{3.11}$$

In Eq. (3.11), P is the pressure, V is the volume, N_{mol} is the number of moles present, T is the absolute temperature, and R is the ideal gas constant. Its value is $R = 8.314\,\mathrm{J/mol/K} = 8.314 \times 10^7 \mathrm{erg/mol/K}$. It is also useful to express the gas law in terms of the mass per unit volume. Let $\rho_{mol} = N_{mol}/V$ be the number of moles/volume. Then if the gas molecules have molecular weight M_{mol}, the mass/volume is $\rho_m = n_{mol}M_{mol}$, and the ideal gas law takes the form

$$P = \rho_m \frac{RT}{M_{mol}}. \tag{3.12}$$

For the atmosphere, $M_{mol} = 28.965\,\mathrm{g}$, very close to the molecular weight of the N_2 molecule, 28.014 g. The mass density of air at sea level and $T = 15°\mathrm{C}$ is $\rho_m = 1.225 \times 10^{-3}\mathrm{g/c}^3$.

Internal Energy of Ideal Gas The internal energy of an ideal gas arises solely from the kinetic energy of the molecules, which in general can include rotational and vibrational kinetic energy in addition to translational kinetic energy. For a monatomic gas with only translational degrees of freedom, the internal energy is

$$U = N_{mol}(\frac{3}{2}RT). \tag{3.13}$$

According to classical thermodynamics, each degree of freedom contributes $RT/2$ to the internal energy. For a monatomic gas, there are three degrees of freedom, one for each direction of motion. This explains the factor 3/2 in Eq. (3.13). The atmosphere is predominantly composed of diatomic molecules such as N_2 and O_2. For such gases at the temperatures in the atmosphere, rotational kinetic energy must be taken into account. In addition to their three translational degrees of freedom, diatomic molecules have two rotational degrees of freedom,[2] each of which contributes an additional $RT/2$ per mole to the kinetic energy. For diatomic molecules and air in particular, we then have

$$U = N_{mol}(\frac{5}{2}RT). \tag{3.14}$$

Specific Heats of an Ideal Gas The specific heat of any system is a measure of how much the internal energy changes per degree of temperature change. There are two specific

[2] Visualizing a diatomic molecule as a dumbbell, the axis of the dumbell can be rotated in two directions.

heats for a gas, denoted as C_V and C_P, according to whether the temperature is changed holding volume V or pressure P constant.

For an ideal gas, the internal energy is given by

$$U = C_V N_{mol} T. \tag{3.15}$$

For the case where heat is added to the system at constant volume, the first law reduces to $dU = TdS = dQ$. The specific heat/mole at constant volume is, by definition,

$$C_V = \frac{1}{N_{mol}}(\frac{dQ}{dT})_V, \tag{3.16}$$

and using $dQ = dU$ along with Eq. (3.15) the specific heat at constant volume is

$$C_V = \frac{1}{N_{mol}}(\frac{\partial U}{\partial T})_V. \tag{3.17}$$

So for a monatomic ideal gas, we have

$$(C_V)_{mono} = \frac{3}{2}R, \tag{3.18}$$

while for a diatomic ideal gas, we have

$$(C_V)_{diatomic} = \frac{5}{2}R. \tag{3.19}$$

Experimental values of specific heats are usually given in J/kg/K. In these units, the value of C_V for dry air at standard temperature and pressure is 717.1J/kg/K. Dividing $5/2R$ by the molecular weight of air (28.965×10^{-3} kg) gives 717.6J/kg/K, so air in the atmosphere is behaving as an ideal gas.

The specific heat/mole at constant pressure for an ideal gas is related to the specific heat/mole at constant volume by the simple formula

$$C_P = C_V + R, \tag{3.20}$$

so for air with $C_V = 5R/2$, we have $C_P = 7R/2$. Eq. (3.20) is easy to derive. The first law now reads

$$dU = TdS - PdV. \tag{3.21}$$

The ideal gas law states

$$P = \frac{N_{mol}RT}{V}, \tag{3.22}$$

so if $dP = 0$, we have $d(T/V) = 0$, which is equivalent to $dT/T = dV/V$. Using this, $PdV = N_{mol}RdT$, so we have

$$(dQ)_P = (TdS)_P = dU + N_{mol}RdT \tag{3.23}$$

$$= N_{mol}C_PdT = N_{mol}(C_VdT + RdT),$$

which leads to

$$C_P = C_V + R. \tag{3.24}$$

For an ideal diatomic gas, the ratio C_P/C_V is 7/5. The value of this ratio for dry air at standard conditions is in close agreement with 7/5, another indication that the atmosphere is behaving as an ideal gas.

Effects of Gravity on the Atmosphere The average properties of the atmosphere as a function of altitude are remarkably stable despite the attractive force of gravity on the air molecules. The reason is that the pressure in the atmosphere is such that a given slab of air is in mechanical equilibrium. If z is the altitude, measured from the surface of the Earth, the equation governing the variation of the pressure is

$$\frac{dP}{dz} = -\rho_m g, \tag{3.25}$$

where g is the gravitational acceleration. This equation, along with the ideal gas law, gives us two equations in the three unknowns of pressure, temperature, and density. To solve for all three, another equation is needed. The simplest possibility, one we will explore in the next section, is to assume the temperature of the atmosphere is constant as a function of altitude. More realistic results are obtained by assuming the atmosphere is adiabatic, a behavior that will be explained after our treatment of the isothermal atmosphere.

Isothermal Atmosphere At first sight, the assumption of a constant temperature for the atmosphere seems reasonable. For example, an observer is more likely to notice "thin air" (reduced density) than any change in temperature when traveling 2 km above sea level. But the decrease of temperature with altitude does become very clear when the altitude increases by several kilometers. By an altitude of 9 km the external temperature is approximately 60°C below the sea level temperature.

Although ultimately unrealistic, the isothermal atmosphere is a useful preliminary to the case of an adiabatic atmosphere. Proceeding, we simplify notation by having $\rho_m \to \rho$, $M_{mol} \to M, N_{mol} \to N$, and, we denote sea level quantities by a subscript 0; P_0, T_0, and ρ_0.

For an isothermal process connecting states 0 and 1, for an ideal gas we have

$$P_1 V_1 = P_0 V_0, \tag{3.26}$$

which is equivalent to

$$P_1 \rho_1^{-1} = P_0 \rho_0^{-1}. \tag{3.27}$$

Applying this to the atmosphere, we take state 1 to be the state of the atmosphere at altitude z, while state 0 is the state of the atmosphere at sea level. Eq. (3.27) becomes

$$P(z)(\rho(z))^{-1} = P_0 \rho_0^{-1}. \tag{3.28}$$

Using this equation to solve for $\rho(z)$ and substituting the result in Eq. (3.25) gives

$$\frac{dP(z)}{dz} = -Mg\frac{\rho_0}{P_0}P(z). \tag{3.29}$$

The ideal gas law at sea level is $P_0 = \rho_0 RT_0$, so Eq. (3.29) becomes

$$\frac{dP}{dz} = -(\frac{Mg}{RT_0})P. \tag{3.30}$$

The combination RT_0/Mg is a characteristic length D_0,

$$D_0 \equiv RT_0/Mg. \tag{3.31}$$

Taking $T_0 = 288\,\mathrm{K}$, Eq. (3.31) gives $D_0 = 8.43\,\mathrm{km}$.

Eq. (3.30) now reads

$$\frac{dP}{dz} = -\frac{1}{D_0}P, \tag{3.32}$$

so that

$$P(z) = P_0 \exp(-\frac{z}{D_0}). \tag{3.33}$$

As long as the temperature is constant, the ideal gas law says that the density and pressure vary with altitude in the same way, so for the density, we have

$$\rho(z) = \rho_0 \exp(-\frac{z}{D_0}). \tag{3.34}$$

To summarize, both pressure and density fall exponentially for the case of an isothermal atmosphere while the temperature remains at the sea level value. Comparing $P(z)$ calculated from Eq. (3.33) to actual data at various altitudes, one finds that Eq. (3.33) is in good agreement with the data up to $z \sim 1\,\mathrm{km}$, but eventually decreases too fast. For example, at $z = 9\,\mathrm{km}$, the pressure calculated from Eq. (3.33) is $\sim 12\%$ smaller than the actual pressure at that altitude.

Adiabatic Atmosphere One reason for favoring an adiabatic description of the atmosphere is the low heat conductivity of air; the heat conductivity of air is $\sim 1/25$ that of water, and $\sim 1/10,000$ that of copper. The other reason is that the atmosphere is rarely, if ever, absolutely still. There are continuous exchanges of parcels of air over various distance and time scales. The heat conductivity of air is so low that these exchanges are essentially adiabatic; a negligible amount of heat is transferred when two small volumes are exchanged.

If states 0 and 1 of an ideal gas are connected by an adiabatic expansion or compression, their pressures and volumes are related by

$$P_1 V_1^{\gamma} = P_0 V_0^{\gamma}. \tag{3.35}$$

In Eq. (3.35), γ is the ratio of the specific heat/mole at constant pressure to the specific heat/mole at constant volume.[3] For diatomic molecules such as N_2 and O_2, we have

$$C_V = \frac{5}{2}R,\ C_P = \frac{7}{2}R, \tag{3.36}$$

so $\gamma = 7/5$. To derive Eq. (3.35), we note that there is no heat transfer in an adiabatic process so $dS = 0$. For a general process, dS is

$$dS = \frac{1}{T}(dU + PdV). \tag{3.37}$$

For an ideal gas, this expression can be transformed into

$$dS = N[C_P \frac{dV}{V} + C_V \frac{dP}{P}]. \tag{3.38}$$

Setting $dS = 0$, we have

$$C_P \frac{dV}{V} + C_V \frac{dP}{P} = 0, \tag{3.39}$$

[3]Fermi uses κ instead of γ.

which is equivalent to the usual form of the adiabatic law,

$$PV^{\gamma} = \text{const.} \tag{3.40}$$

Adapting Eq. (3.35) to the atmosphere, we can relate the pressure and density at altitude z to their sea level values. Eq. (3.35) is equivalent to

$$P(z)\rho(z)^{-\gamma} = P_0\rho_0^{-\gamma}. \tag{3.41}$$

Solving for $\rho(z)$,

$$\rho(z) = \rho_0(\frac{P(z)}{P_0})^{1/\gamma}. \tag{3.42}$$

Using this expression for ρ, Eq. (3.25) now involves only the pressure P and the altitude z. Skipping some elementary steps, the solution for the pressure is

$$P(z) = P_0(1 - (\frac{\gamma - 1}{\gamma})\frac{\rho_0 g z}{P_0})^{\frac{\gamma}{\gamma - 1}}. \tag{3.43}$$

Since $\gamma = 7/5$, we have

$$\frac{\gamma - 1}{\gamma} = \frac{2}{7}.$$

Using the ideal gas law, we have

$$\frac{\rho_0 g}{P_0} = \frac{Mg}{RT_0} = \frac{1}{D_0}. \tag{3.44}$$

Our expression for the pressure at altitude z in an adiabatic atmosphere becomes

$$P(z) = P_0(1 - \frac{2}{7}\frac{z}{D_0})^{\frac{7}{2}}. \tag{3.45}$$

Having found the pressure as a function of altitude, let us find the temperature. First, we use the ideal gas law to solve for the density. This gives

$$\rho(z) = \frac{P(z)M}{RT(z)}. \tag{3.46}$$

Substituting for ρ in Eq. (3.41) and canceling constant factors, we arrive at

$$\frac{P(z)^{1-\gamma}}{T(z)^{-\gamma}} = \frac{P_0^{1-\gamma}}{T_0^{-\gamma}}, \tag{3.47}$$

or

$$\frac{T(z)}{T_0} = (\frac{P(z)}{P_0})^{\frac{\gamma-1}{\gamma}}. \tag{3.48}$$

Using Eq. (3.45) to substitute for $P(z)/P_0$, we finally obtain

$$T(z) = T_0((1 - (\frac{\gamma-1}{\gamma})\frac{\rho_0 g z}{P_0}). \tag{3.49}$$

Substituting for γ and D_0, we have

$$T(z) = T_0(1 - \frac{2}{7}\frac{z}{D_0}). \tag{3.50}$$

We see that the temperature decreases linearly with altitude in an adiabatic atmosphere. At first sight it is worrisome that Eqs. (3.45) and (3.50) predict zeroes in pressure and temperature at $z = 7D_0/2$. However, this altitude is well above the troposphere, and into the next layer of the atmosphere, the stratosphere, where additional factors come into play.

Lapse Rate Remaining in the troposphere, Eq. (3.50) predicts a definite value for the "lapse rate," Γ, which is

$$\Gamma \equiv \frac{dT}{dz}. \tag{3.51}$$

Using Eqs. (3.31) and (3.50), we have

$$\Gamma = -(\frac{\gamma - 1}{\gamma})\frac{Mg}{R}. \tag{3.52}$$

Using the definitions of C_P, C_V, and γ, Γ becomes the very simple expression,

$$\Gamma = -\frac{Mg}{C_P}. \tag{3.53}$$

This expression for Γ, which is known as the "dry air lapse rate," is the ratio of the force of gravity on a mole of air, divided by the specific heat/mole at constant pressure. Its numerical value is -9.8 K/km, so an increase in altitude of 1 km will be accompanied by a decrease in temperature of 9.8 K. While this is accurate for dry air, the actual air in the atmosphere is rarely completely dry. The presence of water vapor in the atmosphere has a large effect on the lapse rate, generally reducing it from the dry air lapse rate to around 6 K/km. The treatment of an adiabatic atmosphere containing water vapor is a more complex subject which is treated in Sec. (3.7). The lapse rate is closely related to D_0. From Eqs. (3.31) and (3.53), we have

$$\Gamma = \frac{T_0}{\gamma D_0}. \tag{3.54}$$

From the adiabatic formula for pressure, Eq. (3.45), we have $P/P_0 = 0.58$ at $z = D_0/2$, and $P/P_0 = 0.31$ at $z = D_0$. At smaller values of z, the adiabatic formula approaches the isothermal formula for pressure. Again, the adiabatic formula for pressure, as expressed in Eq. (3.45), is a dry air formula. The atmospheres over a desert or a snowfield are examples of places where it would hold.

3.5 Chemical Potential. Thermodynamic Potentials

Chemical Potential The cases discussed so far are those of a gas with a fixed number of particles. Thermodynamics applies to the more general case when the number of particles is varied. The corresponding intensive variable is called the "chemical potential" and is usually denoted as μ. If more than one type of molecule is present, each type i has its own chemical potential, μ_i. In the atmosphere there are separate chemical potentials for N_2, O_2, H_2O, etc. The chemical potential plays a central role when particle exchange is present. For the physics of the atmosphere, the transition of water molecules from liquid to vapor forms is the most important example.

Thermodynamic Potentials We now have all the thermodynamic state variables needed to express the first law of thermodynamics in its most general form:

$$dU = TdS - PdV + \sum_i \mu_i dN_i. \tag{3.55}$$

This says that the small change in the internal energy of the gas is the sum of the heat absorbed, minus the work done, plus the extra energy derived from adding particles. In what follows, N_i is the number of moles of molecule i, so μ_i is the chemical potential/mole for species i. Eq. (3.55) is a true differential, meaning that the internal energy is a function of the entropy, the volume, and the N_i, with

$$(\frac{\partial U}{\partial S})_{V,N_i} = T,\ (\frac{\partial U}{\partial V})_{S,N_i} = -P,\ (\frac{\partial U}{\partial N_i})_{S,V} = \mu_i. \tag{3.56}$$

An important result (Callen 1960) for the internal energy is the Gibbs-Duhem relation, which is an explicit formula for the internal energy:

$$U = TS - PV + \sum_i \mu_i N_i. \tag{3.57}$$

Eq. (3.57) follows from the extensive property of the internal energy.

If the internal energy is known as a function of S, V, and the N_i, complete thermodynamic information is available. However, this form is inconvenient for relating to measurable quantities. Temperature and pressure are preferable to their extensive counterparts, entropy and volume. The two forms of "free energy" known as the Helmholtz free energy and the Gibbs free energy are quantities closely related to the internal energy, in which one or more extensive variables have been traded for their corresponding intensive variables.

Consider first the Helmholtz free energy, defined by

$$F \equiv U - TS. \tag{3.58}$$

It is easy to show (Callen 1960) that F is a function of temperature, pressure and the N_i. In terms of $F(T, V, N_i)$, the first law takes the form

$$dF = -SdT - PdV + \sum_i \mu_i dN_i. \tag{3.59}$$

For a process which takes place at constant temperature and constant N_i, the Helmholtz free energy changes only when the system does work or is worked upon, so at constant temperature and N_i, F is independent of the flow of heat in and out of the system. From the Gibbs-Duhem relation, Eq. (3.57) and Eq. (3.58), F is given in explicit form as

$$F = -PV + \sum_i \mu_i N_i. \tag{3.60}$$

The Gibbs free energy takes the process of trading extensive variables for intensive variable one step further. It is defined by

$$G = U - TS + PV = F + PV. \tag{3.61}$$

Again using the Gibbs-Duhem relation, G is given explicitly by

$$G = \sum_i \mu_i N_i. \tag{3.62}$$

The Gibbs free energy is a function of T, P, and the N_i. The first law, when expressed in terms of G, is

$$dG = -SdT + VdP + \sum_i \mu_i dN_i. \tag{3.63}$$

For a process at constant temperature and pressure, the Gibbs free energy changes only when there is particle exchange. At constant T and P, the Gibbs free energy is independent of both the heat flowing in and out of the system and the work done by or on the system. The Gibbs free energy is particularly useful in cases where a gas is in equilibrium with its liquid form, the most important example being water vapor in equilibrium with its liquid form in the atmosphere.

Let us return to the case where there is only one type of molecule present. Then Eq. (3.62) reduces to

$$G = \mu N. \tag{3.64}$$

Defining the Gibbs free energy per mole as $g \equiv G/N$, we see that g is just the chemical potential, μ,

$$g(T, P) = \frac{G}{N} = \mu(T, P). \tag{3.65}$$

Now

$$dG = -SdT + VdP + gdN = dgN + gdN. \tag{3.66}$$

Demanding that the terms on the right side of Eq. (3.66) agree leads to

$$\frac{\partial g(T,P)}{\partial T} = -\frac{S}{N} \equiv -s, \tag{3.67}$$

$$\frac{\partial g(T,P)}{\partial P} = \frac{V}{N} \equiv v, \tag{3.68}$$

so the derivatives of the Gibbs free energy per mole determine the entropy and volume per mole.

3.6 The Boundary between Phases

A given substance can exist in different forms or *phases*, the most familiar example being water, which can be in gaseous, liquid, or solid form. In this section, we will first treat the generic case of a liquid in mechanical and thermal equilibrium with its vapor, and then go on to the particular application considered by Fermi of liquid water in equilibrium with its vapor. We denote the liquid as system 1 and the vapor as system 2. Thermal equilibrium between 1 and 2 requires that $T_1 = T_2$, while mechanical equilibrium demands that $P_1 = P_2$, so we have a common temperature and pressure denoted from now on by T and P. Similarly the chemical potentials of the two phases must be equal since an inequality between μ_1 and μ_2 would lead to a net flow of particles from the system of higher chemical potential into the one of lower chemical potential. Thus the system is only in complete equilibrium when the three intensive variables, T, P, μ, take the same values in the two systems.

It is important to realize that although the chemical potential is the same throughout the entire system, $g_1(T,P)$ and $g_2(T,P)$ are different functions, describing respectively the thermodynamic properties of the vapor and of the liquid. This point is illustrated by considering the volume per mole in the two phases. The vapor is certainly less dense than the liquid, so $v_1 > v_2$. Using Eq. (3.68), we have

$$v_1 = \frac{\partial g_1(T,P)}{\partial P} > v_2 = \frac{\partial g_2(T,P)}{\partial P}. \tag{3.69}$$

It is clear from Eq. (3.69) that $g_1(T,P)$ and $g_2(T,P)$ must be different functions. The same conclusion follows from considering the entropy per mole. From a geometric point of view, we can visualize either g_1 or g_2 as a surface, g_1 and g_2 being the heights above the T,P plane. The physical requirement that $g_1 = g_2$ can only be realized along the curve where the surfaces intersect. This means that for a vapor-liquid system to be in equilibrium, only one of T and P can be chosen freely so that if T is chosen, requiring that $g_1(T,P) = g_2(T,P)$ will determine the pressure. An equation for the coexistence curve follows from writing out $dg_1(T,P) = dg_2(T,P)$:

$$\begin{aligned} dg_1 &= \frac{\partial g_1(T,P)}{\partial T}dT + \frac{\partial g_1(T,P)}{\partial P}dP = \\ dg2 &= \frac{\partial g_2(T,P)}{\partial T}dT + \frac{\partial g_2(T,P)}{\partial P}dP \end{aligned} \tag{3.70}$$

Using Eqs. (3.67) and (3.68), Eq. (3.70) is equivalent to

$$s_1 dT + v_1 dP = s_2 dT + v_2 dP. \tag{3.71}$$

Solving for the derivative along the coexistence curve, we have

$$\frac{dP}{dT} = \frac{s_1 - s_2}{v_1 - v_2}. \tag{3.72}$$

This equation can be put into a more useful form by recasting it in terms of s_g and s_l, the entropies per unit mass of the vapor and liquid, respectively, and likewise v_g and v_l, the volumes per unit mass of the vapor and liquid. If the molecules in the system have molar mass M, then $s_l = s_1/M$, $s_g = s_2/M$, and $v_l = v_1/M$, $v_g = v_2/M$. Using this notation, Eq. (3.72) becomes

$$\frac{dP}{dT} = \frac{s_g - s_l}{v_g - v_l}. \tag{3.73}$$

The heat required to vaporize a unit mass of liquid is called the latent heat[4] and is denoted by Fermi as λ. It is defined by

$$\lambda \equiv T(s_g - s_l). \tag{3.74}$$

In terms of the latent heat, Eq. (3.73) becomes

$$\frac{dP}{dT} = \frac{\lambda}{T(v_g - v_l)}. \tag{3.75}$$

Eq. (3.75) is usually known as the Clausius-Clapeyron equation, or sometimes simply as the Clapeyron equation, Clapeyron's work having predated that of Clausius by a number of years.

Fig. (3.1) shows a portion of the pressure vs. temperature coexistence curve between liquid water and its vapor. It is clear from the graph that lower pressure means lower boiling temperature, a fact familiar to mountain dwellers. It should also be mentioned that if the graph were extended to much higher temperatures and pressures, the coexistence curve would end in what is known as a "critical point," $T_c = 374°\text{C}, P_c = 218\,\text{atm}$. By taking the system on a continuous path around this point, liquid water can be transformed into water vapor with no latent heat.

Assuming the vapor obeys the ideal gas law, we have

$$Pv_g = \frac{RT}{M}, \tag{3.76}$$

[4]The latent heat is denoted as L by other authors.

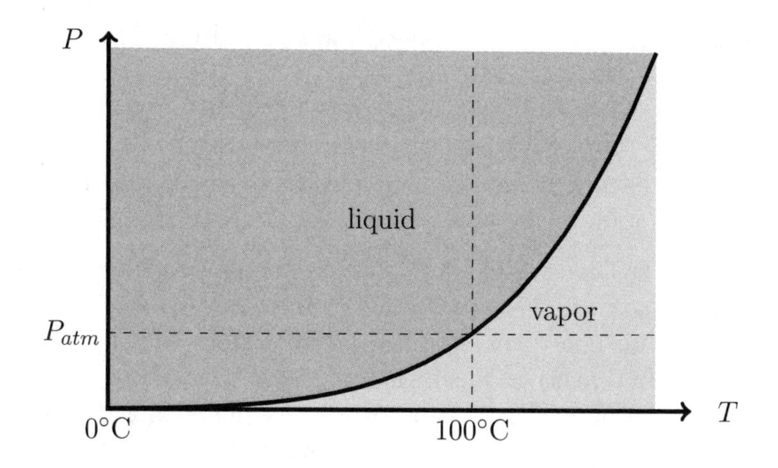

Figure 3.1: P vs. T for water and its vapor

where M is the mass of one mole (18 g for water). Since vapors are generally much less dense than liquids, it is reasonable to ignore v_l compared to v_g in Eq. (3.75). This gives

$$\frac{dP}{dT} \simeq \frac{\lambda}{T(v_g)}. \tag{3.77}$$

Eliminating v_g using Eq. (3.76), we obtain

$$\frac{dP}{dT} = (\frac{\lambda M}{R})\frac{PT}{T}, \tag{3.78}$$

or

$$\frac{dP}{P} = (\frac{\lambda M}{RT})\frac{dT}{T}. \tag{3.79}$$

We can integrate this equation if we ignore the rather modest variation with temperature of the latent heat. Specializing to the case of water, we will take the latent heat to be constant and equal to its value at 100°C, $\lambda_w = 540\,\text{cal/g} = 2,260\,\text{J/g}$. Doing this, and adjusting the constant of integration so that atmospheric pressure is obtained[5] at $T = 100°\text{C} = 373\,\text{K}$, we find

$$P(T) = P_{atm} \exp(13.1 - \frac{4,893}{T}), \tag{3.80}$$

where T is in kelvins. This simple formula is a reasonable approximation to the accurate data plotted in Fig. (3.1).

Since the Clausius-Clapeyron equation applies generally to phase transitions, it is interesting to also consider its prediction for the phase change from ice to water. Whereas we saw earlier that $v_2 \gg v_1$, in this case $v_2 = 1.00\,\text{cm}^3/\text{g}$ is less then $v_1 \approx 1.09\,\text{cm}^3/\text{g}$. In other words, ice floats on water, an anomaly because the solid phase is usually denser than the liquid phase. Moreover, since the difference in volume per gram is orders of magnitude smaller than it is for the vapor to liquid transition, dP/dT is orders of magnitude larger and conversely dT/dP is orders of magnitude smaller.

Even though dT/dP is relatively small for the water-ice transition, it can play an important role in glacier flow since ice encountering a rock-like obstacle experiences a consequent increase in pressure. It may then melt because the negative sign of $(v_2 - v_1)$ implies that dT/dP is also negative. The ice will refreeze almost immediately, but its temporary melting may aid the glacier to slide around the obstacle.

3.7 Adiabatic Transformation of Moist Air

An adiabatic transformation for dry air obeys Eq. (3.40). Taking the ratio of specific heats for air to be 7/5, and using the

[5]Kelvins means temperature with respect to absolute zero. The relation between Celsius temperature T_C and and Kelvin temperature T_K is $T_K = T_C + 273$.

the ideal gas law, the following three quantities are constant in a dry-air adiabatic process:

$$PV^{7/5}, \; VT^{5/2}, \text{ and } PT^{-7/2}.$$

Taking differentials gives

$$\frac{V}{P}\frac{dP}{dV} = -\frac{7}{5}, \; \frac{T}{V}\frac{dV}{dT} = -\frac{5}{2}, \text{ and } \frac{T}{P}\frac{dP}{dT} = \frac{7}{2}. \tag{3.81}$$

In the presence of moisture, all three of these differential relations will be modified. In this section of his notes, Fermi focuses on the last one, involving dP/dT. Before presenting the way in which that relation is modified, it is certainly in Fermi's spirit to ask what is the qualitative expectation; does 7/2 get replaced by a larger or a smaller factor? To answer this, consider an adiabatic expansion of a dry ideal gas. Both the pressure and temperature decrease. Now if moisture is also present in form of water vapor, lowering the pressure slightly will generally lead to a small amount of condensation, and the release of a small amount of latent heat into the surrounding gas. This will make the decrease in temperature, dT, smaller in magnitude than it was before. Looking back at the last relation in Eq. (3.81), we then expect 7/2 to be replaced by a larger number.

The formula derived by Fermi involves two dimensionless ratios which we give the symbols α and β. If λ is the latent heat of water, and M_v is the molecular weight of water, then

$$\alpha \equiv \frac{\lambda M_v}{RT}. \tag{3.82}$$

This combination first appeared in discussing the Clausius-Clapeyron equation. See Eq. (3.79). If P_v is the vapor pressure of water at the chosen conditions, and P is the total pressure, then

$$\beta \equiv \frac{P_v}{P}. \tag{3.83}$$

As a numerical example, suppose the temperature is 20°C = 293 K, a typical temperature for a humid summer day. Then

$$\alpha = \frac{(2260\,\mathrm{J/g})\cdot(18\,\mathrm{g})}{(8.314\,\mathrm{J/mol\,K})\cdot(293\,\mathrm{K})} = 16.7. \tag{3.84}$$

For β, the partial pressure of water vapor at 20°C is 17.5 mmHg. Dividing by atmospheric pressure of 760 mmHg we find

$$\beta = \frac{17.5}{760} = 2.3\times 10^{-2}. \tag{3.85}$$

So β is small, and it is valid to ignore β compared to 1. Fermi also makes this approximation. The parameter α is considerably larger than 1, but it always appears as either $\alpha\beta$ or $\alpha^2\beta$. For the conditions just stated, $\alpha\beta = 0.38$, and $\alpha^2\beta = 6.4$, so $\alpha^2\beta$ is very significant and in fact dominates the final result.

Fermi's formula,[6] with $\alpha^2\beta = 0.38$ and $\alpha\beta = 16.7$, gives

$$\frac{T}{P}\frac{dP}{dT} = \frac{7}{2}\left(\frac{1+\frac{2}{7}\alpha^2\beta}{1+\alpha\beta}\right) = 7.17, \tag{3.86}$$

so the dry air result of 7/2 is indeed increased by condensation of water vapor, by a rather large factor!

Derivation of Fermi's Formula We are interested in an adiabatic process in moist air; dry air plus water vapor. What is adiabatic for the moist air as a whole is not adiabatic for the dry air. The dry air picks up latent heat when water vapor condenses (or gives it up when liquid water evaporates). Applying the first law to the dry air provides a direct route to

[6] Fermi makes a rare algebraic error on page 4B of his notes, inverting numerator and denominator on the right-hand side of Eq. (3.86).

Fermi's formula. We will use a subscript a for dry air quantities, and a subscript v for water vapor quantities. Writing the first law in the form given in Eq. (3.5) for dry air gives

$$dU_a = TdS_a - P_a dV. \tag{3.87}$$

Dry air is treated as an ideal gas which obtains its heat input or output solely from vapor condensation or evaporation. Explicitly, Eq. (3.87) becomes

$$\frac{5}{2}N_a RdT = -\lambda M_v dN_v - \frac{N_a RT}{V}dV, \tag{3.88}$$

where N_a and N_v are mole numbers for dry air and vapor. It is worth remarking on the sign of the first term. Suppose there is condensation, so $dN_v < 0$. The minus sign in front of this term ensures that the resulting latent heat is *added* to the dry air when vapor condenses. Dividing Eq. (3.88) by $N_a RT$ gives

$$\frac{5}{2}\frac{dT}{T} = -\frac{\lambda M_v}{RT}\frac{dN_v}{N_a} - \frac{dV}{V}. \tag{3.89}$$

Since the number of dry air moles is not changing,

$$dN_v/N_a = d(N_v/N_a).$$

Further, since both dry air and vapor are assumed to be ideal gases, we have $d(N_v/N_a) = d(P_v/P_a)$. Rearranging the terms in Eq. (3.89) results in

$$\begin{aligned} \frac{5}{2}\frac{dT}{T} + \frac{dV}{V} &= -\frac{\lambda M_v}{RT} d(\frac{P_v}{P_a}) \qquad (3.90) \\ &= -\frac{\lambda M_v}{RT}\left(\frac{1}{P_a}\frac{dP_v}{dT}dT + \frac{P_v}{P_a}(\frac{dV}{V} - \frac{dT}{T})\right), \end{aligned}$$

where we use the ideal gas law to evaluate $d(1/P_a)$. We assume the moist air is *saturated* so the derivative of P_v can be evaluated using the Clausius-Clapeyron equation, Eq. (3.79). This results in

$$\frac{1}{P_a}\frac{dP_v}{dT}dT = (\frac{\lambda M_v}{RT})\frac{P_v}{P_a}\frac{dT}{T}. \tag{3.91}$$

Returning to Eq. (3.90), it now reads

$$\frac{5}{2}\frac{dT}{T} + \frac{dV}{V} = -(\frac{\lambda M_v}{RT})\frac{P_v}{P_a}\left((\frac{\lambda M_v}{RT})\frac{dT}{T} + \frac{dV}{V} - \frac{dT}{T}\right). \tag{3.92}$$

At this point it is convenient to express things in terms of α and β. The total pressure of the moist air is $P = P_a + P_v = P_a + \beta P$, so $P_a = (1 - \beta)P$. Since β is quite small, it is a good approximation to follow Fermi and replace P_a with P; $P_v/P_a \approx P_v/P$. Doing so, we get

$$\frac{5}{2}\frac{dT}{T} + \frac{dV}{V} = -\alpha\beta\left(\alpha\frac{dT}{T} + \frac{dV}{V} - \frac{dT}{T}\right). \tag{3.93}$$

Rearranging this equation, we have

$$\left(\frac{5}{2} + \alpha\beta(\alpha - 1)\right)\frac{dT}{T} + (1 + \alpha\beta)\frac{dV}{V} = 0. \tag{3.94}$$

For an ideal gas, we have

$$\frac{dP}{P} + \frac{dV}{V} = \frac{dT}{T}. \tag{3.95}$$

Using this formula to replace dV/V in Eq. (3.94) gives

$$(\frac{7}{2} + \alpha^2\beta)\frac{dT}{T} = (1 + \alpha\beta)\frac{dP}{P}, \tag{3.96}$$

or

$$\frac{T}{P}\frac{dP}{dT} = \frac{7}{2}\left(\frac{1 + \frac{2}{7}\alpha^2\beta}{1 + \alpha\beta}\right), \tag{3.97}$$

which is Fermi's formula.

The dominant effect in the modification of 7/2 to a bigger number is the large latent heat of water. The water vapor plays a small role in the pressure and internal energy of moist air. However, the condensation of vapor releases the latent heat into the surrounding gas and significantly affects the adiabatic law which holds for dry air.

3.8 Atmospheric Composition and Temperature

As mentioned in Sec. (3.4), the dominant components of the lower atmosphere are the diatomic molecules of nitrogen and oxygen, by volume respectively 78.08% and 20.95% of the atmosphere at sea level, 15°C and atmospheric pressure of 101,325 Pa = 1 atm. Argon, at 0.93%, constitutes the majority of the remainder. The fourth-largest constituent, one that has drawn increasing focus because of its effect on global warming, is carbon dioxide, at 0.033%. With the exception of neon, at 0.0018%, no other gas rises to the level of parts per thousand. Nevertheless two other gases in the atmosphere have drawn public attention: methane because of its effect on global warming, and ozone because of its dual effect as a greenhouse gas at levels up to about 10 km, in the troposphere, and as a protector against ultraviolet radiation at higher levels (in the stratosphere). In addition the atmosphere contains a variable amount of water vapor, ranging from trace amounts to as high as 4–5%.

3.9 Temperature as a Function of Altitude

As we see in Fig. (3.2), the temperature of the atmosphere decreases linearly as a function of altitude up to approximately 10 km above sea level. As discussed in Sec. (3.4), linear decrease of temperature with altitude occurs in a dry adiabatic atmosphere, but the rate of decrease in the actual atmosphere is smaller than it would be in dry air. At an altitude of 10 km, the temperature has dropped to approximately −50°C.

Leaving the troposphere one enters the stratosphere, a region in which the ozone molecule (O_3) plays an important role both in heating its surroundings and in protecting the Earth against the harmful effects of solar ultraviolet radiation. The temperature above the troposphere rises slowly at first in the region known as the tropopause and then more rapidly, reaching a peak of 50°C at 50 to 60 km above sea level. The stratosphere is known to be very dry and to have little convection, which in turn implies little turbulence, a distinct advantage for aviation. This stability may, however, be disturbed by volcanic eruptions or thunderstorms. Above the stratosphere the temperature begins once again to drop.

The heating of the stratosphere is largely due to ultraviolet solar radiation operating in a process known as photolysis during which its ozone is rapidly destroyed and recreated in a cyclical manner. It begins with an oxygen molecule (O_2) in the stratosphere being split into two oxygen atoms by the absorption of ultraviolet radiation with a wavelength $\lambda \leq 242\,\text{nm}$. An oxygen atom (O) then combines with an oxygen molecule to form an ozone molecule (O_3): during this process energy is released that is transformed into heat. The ozone molecule is then rapidly disassociated by lower-energy solar radiation ($\lambda \leq 320\,\text{nm}$), and the cycle begins anew. In brief, solar ultraviolet radiation is the source of the heating.

This important layer of ozone, the protector from the harmful effects of ultraviolet radiation, can be partially destroyed by the catalytic action of other molecules in the stratosphere. Chlorofluorocarbons are a well-known example. The action of photolysis on them releases chlorine atoms (Cl) which com-

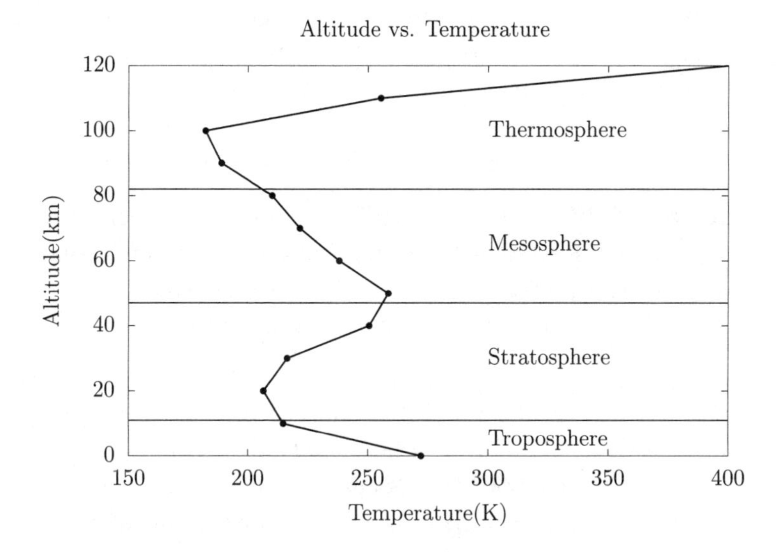

Figure 3.2: Temperature vs. altitude in the Earth's atmosphere.
Source: NASA

bine with ozone (O_3)to form $ClO + O_2$. The ClO molecules then interact with an oxygen atom, $ClO + O \rightarrow Cl + O_2$, and chlorine is free again to bond with ozone. In other words, chlorine is a very effective catalyst for the destruction of ozone.

3.10 Exotic Phenomena in the Upper Atmosphere

Twilight: The Earth's lower atmosphere is illuminated even when the Sun is below the horizon because of sunlight being scattered by dust particles in the upper atmosphere. The phenomenon occurs both near dusk and near dawn.

Luminous Clouds: Also known as noctilucent clouds, these are caused by sunlight being scattered by ice crystals in the mesosphere. They are too faint to be seen other than during twilight.

Zodiacal Light: This is a diffuse light around the Sun caused by the scattering of sunlight by dust in the zodiac or ecliptic, the apparent path in the celestial sphere swept out by the Earth during the course of the year.

Note: There are other notable sources of transitory phenomena of dust in the atmosphere that affect sunlight, such as volcanic eruptions or the passage and subsequent breakup of meteors entering the atmosphere. A discussion of additional phenomena caused by electromagnetic radiation such as auroras is postponed until Chap. (17).

(6)

Loss of atmosphere

$$b^2 = r_0^2 + \frac{2MG}{v^2} r_0$$

b

r_0

Maxwell distribution

$$\sqrt{\frac{54}{\pi}} \frac{v^2 dv}{c^3} e^{-\frac{3v^2}{2c^2}} = p(v)dv \qquad c^2 = \frac{3kT}{m}$$

No of particles lost per unit time

$$\int_0^\infty p(v)dv \frac{\pi b^2}{4\pi r^2} 4\pi r^2 v \; n \qquad n = n_0 e^{-\frac{3MG}{Rc^2}}$$

$$= \sqrt{24\pi}\, n_0 \frac{MGR}{c} e^{-\frac{3MG}{Rc^2}} \left[1 + \frac{Rc^2}{3MG}\right]$$

$GM = gR^2$ assume $\frac{3gR}{c^2} \gg 1$

equiv height of atmosph $= \frac{c^2}{3g}$

life time of atmosphere $= \frac{\sqrt{2\pi}}{3^{3/2}} \frac{c^3}{g^2 R} e^{\frac{3gR}{c^2}}$

τ	c	$T(H)$	$T(H_2)$	$T(He)$	$T(N_2)$
1000 years	2.83×10^5	~~1000~~ 333	~~2000~~ 667	~~4000~~ 1333	~~14000~~ 4670
10^9 years	2.23×10^5	~~625~~ 208	417	834	2410

$$1207\sqrt{\frac{T}{273}} = c_{He}$$

Escape velocity from Earth's atmosphere

CHAPTER 4

Loss of Planetary Atmosphere

4.1 Maxwell Distribution

The starting point of a discussion about particle loss from the atmosphere is the Maxwell-Boltzmann distribution of velocities for the particles of an ideal gas at temperature T. That distribution, with the integral over all velocities normalized to one, is (Mandl 1971)

$$f(\boldsymbol{v})d^3\boldsymbol{v} = \pi^{-3/2}U^{-3}e^{-v^2/U^2}d^3\boldsymbol{v}, \tag{4.1}$$

where

$$U \equiv \sqrt{2kT/m},$$

m is the particle's mass, and k is equal to Boltzmann's constant, a constant that is related to the perfect gas constant R by a factor of Avogadro's number, $R = N_A k$. The meaning of U is that it is the most probable magnitude for the particle velocity. Fermi uses the parameter c, which is the root-mean-square velocity. The two are related by

$$c = \sqrt{\frac{3}{2}} U, \tag{4.2}$$

so Eq. (4.1) for $f(\boldsymbol{v})$ is the equivalent to Fermi's expression. The average kinetic energy of a particle in an ideal gas is expressed in terms of c^2 by

$$(\frac{1}{2}mv^2)_{ave} = \frac{1}{2}mc^2 = \frac{3}{2}kT. \tag{4.3}$$

4.2 Escape Velocity

Gas particles with sufficient velocity may escape from the atmosphere. A naive formulation says this will occur if the overall particle energy, the sum of its kinetic and potential energy, is greater than zero, so the particle is able to overcome the Earth's gravitational pull. This requires

$$mv^2/2 - GM_e m/r \geq 0, \tag{4.4}$$

where G is Newton's constant, M_e is the Earth's mass, r is the particle's distance from the Earth's center, and the particle's velocity is directed radially away from the Earth.

This is overly simplistic, for the gas particles in the high-velocity tail of the Maxwell distribution will in general rescatter and therefore not escape from the atmosphere because their velocity will be redirected many times by rescattering. For example, the mean free path of an air molecule at sea level is approximately 65 nm, so small that in traveling 1 mm, the molecule will undergo over ten thousand collisions. It therefore makes no sense to calculate an "outward flux" at sea level. However, the mean free path of a molecule steadily increases as the density of the atmosphere decreases. At the "top" of the atmosphere, the mean free path is very large and an outgoing molecule has little chance of being scattered before it escapes.

The situation is illustrated in Fig. (4.1), which shows the velocity vector (bold arrow) of an escaping particle at angle θ relative to a vector from the center of the Earth (thin arrow). The outer shaded layer in the figure represents the lower layers of the Earth's atmosphere (troposphere, stratosphere, mesosphere, thermosphere). We define the "top" of the atmosphere to be the lower edge of the exosphere, commonly known as the "exobase," a region located between 500 km and 1000 km above the Earth's surface. The mean free path of a molecule there is thousands of meters. We call r_c the critical distance from the center of the Earth at which the particle is within a mean free path of the "top" of the atmosphere, and T_c the atmosphere's equilibrium temperature at this radius. For definiteness, in the following, we take r_c to correspond to an altitude of 1000 km, and $T_c = 1000\,\mathrm{K}$.

The value of the escape velocity at distance r_c is

$$v_c = \sqrt{2GM_e/r_c}. \tag{4.5}$$

Denoting the density of particles at r_c by n_c, the British astrophysicist Sir James Jeans defined what is called the Jeans particle escape flux, the number of particles expected to escape per second per unit of area (Jeans 1916); (Gross 1974). We denote this quantity by F. The outward flux for particles with given velocity $\boldsymbol{v}$ is proportional to the component of the particle's velocity along the outward direction, given by $n_c \boldsymbol{v} \cdot \hat{\boldsymbol{r}} = n_c v \cos\theta$, where for outgoing particles $\boldsymbol{v} \cdot \hat{\boldsymbol{r}} > 0$, or $0 \le \theta \le 90°$. Multiplying the Boltzmann distribution by this quantity, and integrating over outgoing particles with velocities greater than v_c, we obtain the following expression for

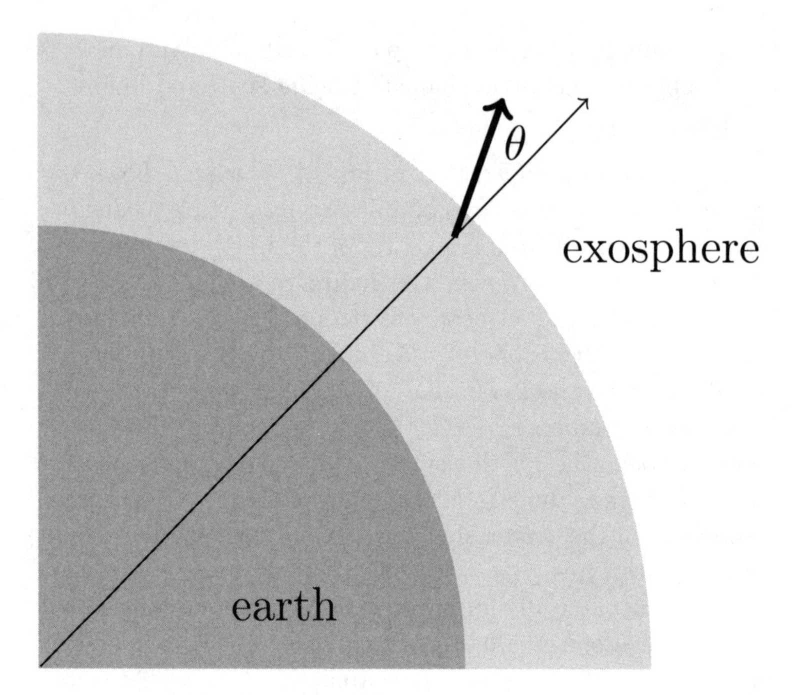

Figure 4.1: Atmospheric escape

F

$$F = \left(\int d^3\boldsymbol{v} f(\boldsymbol{v}) n_c \boldsymbol{v} \cdot \hat{\boldsymbol{r}} \right)_{\boldsymbol{v}\cdot\hat{\boldsymbol{r}}>0, |\boldsymbol{v}|>v_c} . \tag{4.6}$$

Integrating over the angles of the escaping particle yields

$$F = \frac{n_c}{\sqrt{\pi}} \int_{v_c}^{\infty} dv (\frac{v}{U_c})^3 e^{-\frac{v^2}{U_c^2}}, \tag{4.7}$$

where

$$U_c = \sqrt{\frac{2kT_c}{m}}. \tag{4.8}$$

The subscript on U_c means that it is the most probable velocity of a particle of mass m at temperature T_c. The remaining integral over v is simple, as can be seen by using as variable $x = v^2/U_c^2$ and observing that

$$\int_{x_o}^{\infty} xe^{-x}dx = (x_0 + 1)e^{-x_0}. \tag{4.9}$$

We finally obtain

$$F = n_c U_c \left(\frac{1}{2\sqrt{\pi}} (1 + \frac{v_c^2}{U_c^2}) e^{-v_c^2/U_c^2} \right), \tag{4.10}$$

showing that F involves the natural prefactor $n_c U_c$, multiplied by a dimensionless function of $(v_c/U_c)^2$.

Fermi introduces an interesting way of parameterizing $(v_c/U_c)^2$. Begin by defining a local acceleration due to gravity,

$$g_c = \frac{GM_e}{r_c^2}. \tag{4.11}$$

At an altitude of 1000 km, and taking 6371 km for the Earth's radius, the value of g_c is

$$g_c = (\frac{6371}{7371})^2 (9.8\,\mathrm{m/s^2}) = 7.3\,\mathrm{m/s^2}, \tag{4.12}$$

about 75% of g at the Earths's surface. Then using Eq. (4.5) for v_c and the Eq. (4.11) for g_c, we obtain

$$(\frac{v_c}{U_c})^2 = \frac{2GM_e}{r_c}(\frac{1}{U_c})^2 = r_c(\frac{2g_c}{U_c^2}) \equiv \frac{r_c}{h}, \tag{4.13}$$

where the length h is given by

$$h = U_c^2/2g_c = \frac{kT_c}{mg_c}, \tag{4.14}$$

or equivalently,

$$mg_c h = kT_c. \tag{4.15}$$

Now consider the gravitational potential energy of an atom of mass m located at r_c+h. Assuming h is much smaller than r_c, we can work to first order in h/r_c. We have

$$-\frac{GM_e m}{r_c+h} \approx -\frac{GM_e m}{r_c} + m(\frac{GM_e m}{r_c^2})h = -\frac{GM_e m}{r_c} + mg_c h, \tag{4.16}$$

so $mg_c h$ is the potential energy at h of an atom of mass m, relative to its value at r_c.

At an altitude of 1000 km and a temperature of 1000 K, h turns out to be 1132 km for a hydrogen atom. Rewriting Eq. (4.10) in terms of r_c/h, we have

$$F = \frac{U_c n_c}{2\sqrt{\pi}}(1+\frac{r_c}{h})e^{-\frac{r_c}{h}} \approx \frac{U_c n_c}{2\sqrt{\pi}}(\frac{r_c}{h})e^{-\frac{r_c}{h}}. \tag{4.17}$$

From here on, we will use the expression on the extreme right of Eq. (4.17) for F. The exact form of the factor multiplying the exponential in Eq. (4.17) plays an insignificant role compared to the exponential. For hydrogen,

$$r_c/h = (7371/1132) = 6.51. \tag{4.18}$$

The quantities n_c, h, and U_c all vary from atom to atom, so to denote atom a, we write $n_c(a), h(a), U_c(a)$, and F_a. Given the dampening due to the exponential, $h(a)$ can also be interpreted as the effective depth for the layer of the atmosphere from which the particle with mass m_a can be emitted. For an atom a of mass m_a, the value of r_c/h is that of hydrogen multiplied by the ratio of atomic masses, m_a/m_H. Due to the exponential dependence of $F(a)$ on r_c/h, this means that $F(a)$ decreases quite rapidly in going from hydrogen to the heavier atoms in the atmosphere.

4.3 Lifetime of Atmosphere for Various Atoms

Let N_a denote the total number of atoms (or molecules) of type a in the layer of the Earth's atmosphere of thickness $h(a)$. Likewise, dN_a/dt is the rate at which particles of type a are being lost from the atmosphere. It is obtained by multiplying F_a by the area, $4\pi r_c^2$,

$$\frac{dN_a}{dt} = 4\pi r_c^2 F_a. \tag{4.19}$$

We may estimate a lifetime τ_a for species a by the following simple equation;

$$N_a = \frac{dN_a}{dt}\tau_a, \tag{4.20}$$

or

$$\tau_a = N_a(\frac{dN_a}{dt})^{-1}. \tag{4.21}$$

From F_a and Eq. (4.19), we have a definite expression for dN_a/dt. We estimate N_a by multiplying the density $n_c(a)$ by the volume of a spherical shell of thickness $h(a)$,

$$N_a \approx 4\pi r_c^2 h(a) n_c(a). \tag{4.22}$$

Using this approximation, we arrive at a picture of how lifetimes vary as we vary the type of atom. From Eqs. (4.17) and (4.22), we obtain the following expression for τ_a;

$$\tau(a) = 2\sqrt{\pi}\frac{h(a)}{U_c(a)}\left(\frac{h(a)}{r_c}\exp(\frac{r_c}{h(a)})\right) \tag{4.23}$$

$$= \frac{1}{g_c}\sqrt{\frac{2\pi k T_c}{m(a)}}\left(\frac{h(a)}{r_c}\exp(\frac{r_c}{h(a)})\right).$$

To proceed, we use the the same parameters as above; Earth radius of 6371 km, the "top of the atmosphere" altitude of

1000 km, and $r_c = 7371$ km. Taking account of the factor of $1/\sqrt{m(a)}$ in the prefactor and multiplying $r_c/h(a)$ by the ratio $m(a)/m(H)$, it is straightforward to generate Table (4.1). It is evident from the numbers in the table that atoms heavier

Table 4.1: Estimated lifetimes of atmosphere for light atoms

a	$\tau(a)$ year
H	3.3×10^{-3}
H_2	8×10^{-1}
He	1.35×10^{5}
C	1.32×10^{27}

than helium have lifetimes for Jeans escape which are much larger than the age of the Earth. While the escape velocity is independent of the mass of the escaping particle, the needed kinetic energy grows linearly with mass and becomes larger and larger vs. kT_c.

4.4 Magnitude of Particle Losses

As we said earlier, most of the particles escaping will do so from the "top" of the atmosphere, the zone within one mean free path of free streaming. For Earth, this zone is known as the exosphere, its bottom lying at approximately 1000 km above sea level. The temperature there is estimated at roughly 1000 K though values of both T_c and r_c within a factor of two are often considered.

At $r_c = 7371$ km, the escape velocity, as defined in Eq. (4.5), is $v_c = 10.4$ km/s. It is of course the same for all particles, while U_c, the most probable velocity at T_c, is inversely proportional to the square root of the particle's mass. Jeans escape is usually parameterized using the rms velocity U_{rms} (Fermi's c,) related to U_c by $U_{rms} = (\sqrt{3/2})U_c$. A rough rule of thumb (Catling and Zahnle 2009) for estimating Jeans escape loss is that the loss is only significant if

$$\lambda = v_c/U_{rms} \leq 6.$$

At the temperature of 1000 K, we have $U_{rms} = 4.98\,\mathrm{km/s}$ for a hydrogen atom, so $\lambda_H \sim 2$. Consequently we expect the Jeans mechanism to be significant for atomic and molecular hydrogen and possibly even for helium, which has a λ of only 4, still below the limit of 6. On the other hand, the mechanism does not lead to a loss of heavier gases such as oxygen and nitrogen.

In detail atmospheric escape is more complicated than Jeans escape for a number of reasons. One is that the Jeans mechanism is an overestimate of particle loss because it assumes the tail of the Maxwell velocity distribution from which the particles escape is replenished as fast as, or faster than particles escape. This does not seem to be the case. A second reason is the existence of other mechanisms (Gross 1974; Catling and Zahnle 2009) for particle escape that play an important role in hydrogen loss and are dominant for heavier atoms. A detailed discussion of them is outside the scope of Fermi's notes, but we will mention one that is known to play an important role for hydrogen and for helium. A chemical reaction in the upper atmosphere can produce a rapid ion. The Earth's magnetic field normally acts to prevent such an ion from escaping, but the field's lines of force loop back toward the Earth near the poles so the ion may avoid there the constraining force the field imposes. This phenomenon is known as polar wind. Alternatively an ion, e.g., a fast hydrogen ion produced in a chemical reaction, may strip away an electron from a slow hydrogen atom, which now neutral, may proceed without suffering much loss of velocity.

Another effect leading to escape is the solar wind, the stream of charged particles emitted at the Sun's corona. It is largely deflected by the Earth's magnetic field, except near the poles, but it can certainly act in that region to attract and then drag away positively charged ions present in the upper atmosphere. Hydrodynamic escape, in which a strong thermal escape of lighter particles drags along heavier particles by collision is yet another example.

4.5 Helium

We present here a few facts about helium, present at approximately 5 ppm (parts per million) in the atmosphere at standard conditions of pressure and temperature (STP). Since helium is such a light atom, it is at first sight surprising that it exists at all as a component of the lower atmosphere. The explanation is that there is a constant influx into the atmosphere of helium atoms produced in the Earth's crust. They originate through the emission of alpha particles (helium nuclei) in radioactive decay, principally caused by uranium and thorium isotopes. The alpha particles then attach to electrons and form atomic helium.

Helium escapes from the atmosphere at approximately 1.5×10^6 kg/year, (Catling and Zahnle 2009), a much larger rate than the Jeans mechanism predicts. Our present understanding is that whereas the Jeans mechanism accounts for almost half of atmospheric hydrogen's escape, atmospheric helium escape is almost entirely due to other mechanisms, principally polar wind and chemical exchange.

The helium fraction in the atmosphere appears to be stable, although there is no fundamental physics requirement that this be so. A stable fraction would imply that the rate at which helium escapes from the lower atmosphere is balanced by the rate at which helium from radioactive decays is emitted into the lower atmosphere. Neither of these rates is known with great precision although both have been intense subjects of research in the recent years.

Helium has many practical applications: in medicine, computer chip manufacture, high-speed internet cable manufacture, welding, and fundamental physics research, to name just a few. This has led to a helium extraction industry, whose main source is the small fraction of helium in natural gas. The rate at which industrial helium is produced is usually quoted in cubic meters at STP per year. Fermi quotes a figure of $15 \times 10^6\,\mathrm{m^3/year}$ for the amount of helium produced in the United States. The present figure, approximately 40% of the world's total, is closer to $70 \times 10^6\,\mathrm{m^3/year}$. At STP, the density of helium is $0.179\,\mathrm{kg/m^3}$, so the US production of helium is roughly $13 \times 10^6\,\mathrm{kg/year}$. This is several times larger than the amount of helium that escapes the atmosphere in a year. Since helium is a nonrenewable resource, there have been recent concerns about the world's supply of this gas being exhausted. Helium stockpiling, new extraction methods, and helium recycling are all being pursued to ensure a steady supply of helium for the foreseeable future.

4.6 Speed of Sound and Atmospheric Altitude

Fermi briefly illustrates how the speed of sound varies with altitude as sound waves propagate through the atmosphere at different altitudes. Sound waves are caused by successive compressions and rarefactions of the medium through which the disturbance is propagating. The speed of the longitudinal wave therefore depends on the relation between the pressure P that is exerted and the density ρ of the medium. The square of the sound speed is given by

$$v_s{}^2 = (\frac{\partial P}{\partial \rho})_{ad}. \quad (4.24)$$

The subscript *ad* means that the compressions and rarefactions take place adiabatically; pressure and volume are related, as discussed in Sec. (3.4), by

$$P = const.\rho^\gamma.$$

Isaac Newton was the first to evaluate the speed of sound but he erroneously assumed that the successive compressions and rarefactions took place isothermally. Pierre Laplace later realized that the heat flow from one region to another is negligible as long as the wavelength is long compared to the mean free path in the medium. Computing the partial derivative of Eq. (4.24) gives

$$v_s{}^2 = \frac{\gamma P}{\rho}. \quad (4.25)$$

For an ideal gas of molecular weight M, we have

$$\frac{P}{\rho} = \frac{RT}{M}, \quad (4.26)$$

so

$$v_s{}^2 = \frac{\gamma RT}{M}. \quad (4.27)$$

Though not providing the formulas we have just shown, Fermi illustrates graphically how the velocity of sound first decreases with increasing altitude in the atmosphere and then increases as we transition from the troposphere to the stratosphere, following the same pattern of temperature versus atmospheric altitude we have already displayed in Fig. (3.2).

In the troposphere, using $M = 28.97\text{g}$ for the molecular weight of air, $\gamma = 7/5$, and taking $T = 373\,\text{K}$, the speed of sound waves is

$$v_s = \sqrt{\frac{\gamma RT}{M}} = 343\,\text{m/s}. \tag{4.28}$$

The corresponding result for the rms velocity U_{rms} is

$$U_{rms} = \sqrt{\frac{3RT}{M}} = 502\,\text{m/s}. \tag{4.29}$$

We see that the sound velocity is somewhat smaller than the rms velocity. The ratio is independent of temperature and is given by

$$\frac{v_s}{U_m} = \sqrt{\frac{\gamma}{3}} = \sqrt{\frac{7}{15}} = 0.68\,. \tag{4.30}$$

(7)

Relation between vapor pressure and radius of drops

γ = Surface tension = 77 (for Water)

$$\Delta p_{vap} = \rho_{vap}\, h g$$

$$p_{capill} = \frac{2\gamma}{r} = g h \rho_{liq}$$

$$\Delta p_{vap} = \frac{2\gamma}{r} \frac{\rho_{vap}}{\rho_{liq}}$$

use approximate relations for H_2O $\quad \rho_{liq} = 1$

$$\rho_{vap} = \frac{18\, p_{vap}}{RT}$$

$$\frac{\Delta p_{vap}}{p_{vap}} = \frac{2\gamma}{r} \frac{18}{RT}$$

at $T \cong 300$

$$\frac{\Delta p_{vap}}{p_{vap}} = \frac{1.1 \times 10^{-7}}{r}$$

Relation between radius of drops and fall velocity

Small drops: From Stoke's formula (int. friction of air = 17.3×10^{-5})

$$g \times \frac{4\pi}{3} r^3 \times \rho = 6\pi\mu r v$$

$$v = 1.26 \times 10^{6}\, r^2 \text{ cm/sec}$$

Large drops

$$g \times \frac{4\pi}{3} r^3 \times \rho = K \times \pi r^2 \times v^2$$

(K = coeff of air resistance for spherical shape)

$v = 1344\sqrt{r}$ at sea level

$= 2200\sqrt{r}$ at 4 Km height

Diagram for Fermi's calculation of δP_{vap}

CHAPTER 5

Liquid Drop Physics

5.1 Vapor Pressure and Radius of Raindrops

In his discussion of thermodynamics Fermi discussed the problem of phase transitions between a gas and a liquid, and here he extends that discussion by considering the formation of a drop of liquid in a saturated vapor. This requires going beyond the earlier treatment and examining the effect of the surface tension of the drop on the phase transition.

Consider a liquid in equilibrium with its vapor, for example, a closed container containing pure water in both liquid and vaporous forms, with liquid water lying at the bottom of the container and water vapor in the space above. Mechanical equilibrium requires that the pressure in the vapor equal the pressure in the liquid. Denoting liquid and gas phases by subscripts 1 and 2, this states

$$P_1 = P_2. \tag{5.1}$$

This equation applies to the case where the interface between the liquid and vapor is a plane. However, when treating the mechanical equilibrium between a liquid drop and surrounding vapor, there are curved surfaces, and surface tension must be taken into account. Eq. (5.1) is modified. Surface tension has units of force/length and will be denoted in the following by γ. Numerical values of surface tension in SI units are usually given in milli-newtons/meter, equivalent to the cgs unit of dynes/cm. The surface tension of water is approximately

$$\gamma_{wat} = 73\,\mathrm{mN/m}. \tag{5.2}$$

In the case of a droplet surrounded by vapor, surface tension acts to minimize the area of the droplet. The result is that the pressure inside the droplet is larger than the pressure of the surrounding vapor, so $P_1 > P_2$.

An equation for $P_1 - P_2$ is easily obtained using a balance-of-forces treatment, illustrated in Fig. (5.1). Take a spherical droplet of radius r, surrounded by vapor. Divide the sphere exactly in half and consider the forces on the upper hemisphere. In the upward direction the pressure of the liquid exerts a force $\pi r^2 P_1$. In the downward direction, the pressure of the vapor exerts a force of $\pi r^2 P_2$. The surface tension also exerts a downward force on the upper hemisphere. Acting along the circumference of the hemisphere, the surface tension force is $2\pi r\gamma$. Equating upward and downward forces gives

$$\pi r^2 P_1 = \pi r^2 P_2 + 2\pi r\gamma, \text{ or } P_1 - P_2 = \frac{2\gamma}{r}. \tag{5.3}$$

The spherical drop is a special case of a more general formula due to Young and Laplace(Landau and Lifshitz 1980) that involves the two principal radii of curvature of the surface at the point of contact of the two phases. Eq. (5.3) can be used to compute the pressure difference between a raindrop and

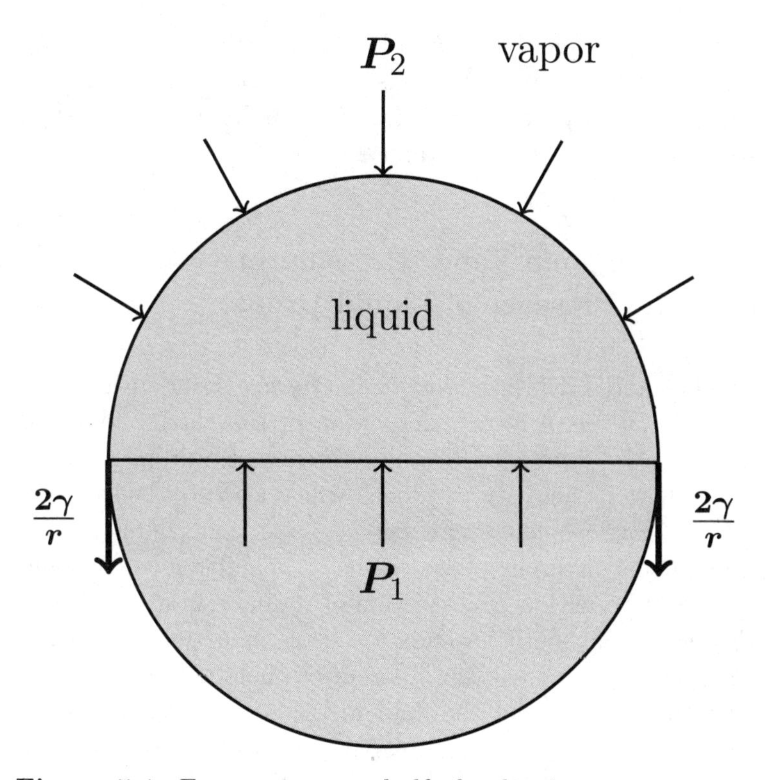

Figure 5.1: Forces on upper half of a droplet

the surrounding water vapor. Taking $\gamma = 73\,\mathrm{mN/m}$, and a raindrop of radius[1] $r = 1\,\mathrm{mm}$, we have

$$\frac{2\gamma}{r} = 146\,\mathrm{Pa}. \tag{5.4}$$

This is tiny compared to atmospheric pressure of $101,325\,\mathrm{Pa}$, so the pressure inside a typical raindrop is only slightly higher than atmospheric pressure.

The presence of a droplet also makes a small change in the vapor pressure of the surrounding vapor, compared to

[1] The range of raindrop radii is roughly $0.1\,\mathrm{mm}$ to $5.0\,\mathrm{mm}$.

its value for a plane interface between liquid and vapor. We follow Fermi's treatment of this in the next section, where we will use subscripts that indicate liquid and vapor rather than the numerical subscripts 1 and 2.

5.2 Change in Vapor Pressure in the Presence of Liquid Drops

Landau and Lifshitz's *Statistical Physics* (Landau and Lifshitz 1980) has a formal derivation of the change in vapor pressure, δP_{vap}, caused by the existence of a drop. Here we follow Fermi's heuristic approach which leads to the same result. Consider a large closed receptacle that contains both a liquid and a vapor phase of the same substance. A long, narrow tube with a cross-section of radius r is suspended in the receptacle; both its ends are kept open (see Fig. (5.2)). Capillary action creates a difference in height h between the fluid in the tube and the fluid in the rest of the container. The small difference in the vapor pressure between what is exerted on the surface of the fluid in the tube and what on the surface of the rest of the container is simply due to the different amounts of vapor above each surface.

Consider first a column of vapor outside the tube. Requiring that this vapor column be in mechanical equilibrium, we have

$$\delta P_{vap} = P_{vap}(0) - P_{vap}(h) = \rho_{vap} g h. \tag{5.5}$$

Turning to the column of liquid in the tube, since the liquid at the bottom of the tube is at the same level as the vapor outside the tube, the pressure at the bottom of the tube is $P_{vap}(0)$. At the top of the tube, the vapor applies the pressure $P_{vap}(h)$. In addition, if we take the contact angle to be 90°, surface tension provides an upward effective pressure

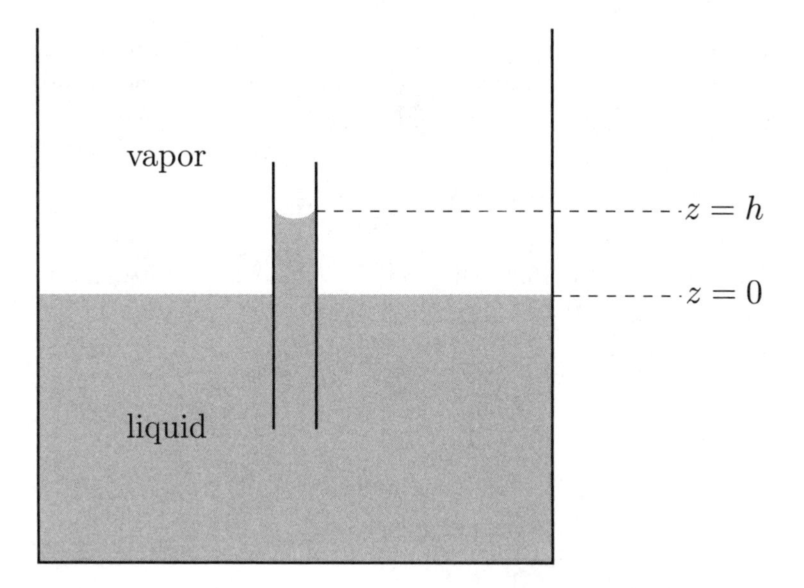

Figure 5.2: Diagram for Fermi's calculation of δP_{vap}

on the liquid of $2\gamma/r$.[2] Requiring the tube of liquid to be in mechanical equilibrium gives

$$P_{vap}(0) + \frac{2\gamma}{r} = P_{vap}(h) + \rho_{liq} gh. \tag{5.6}$$

From Eqs. (5.5) and (5.6) we have two expressions for gh. Setting them equal gives

$$gh = \frac{P_{vap}(0) - P_{vap}(h)}{\rho_{vap}} = \frac{P_{vap}(0) - P_{vap}(h) + \frac{2\gamma}{r}}{\rho_{liq}}. \tag{5.7}$$

Solving for $P_{vap}(0) - P_{vap}(h) = \delta P_{vap}$, we have

$$\delta P_{vap} = \frac{\rho_{vap}}{\rho_{liq} - \rho_{vap}} (\frac{2\gamma}{r}) \approx \frac{\rho_{vap}}{\rho_{liq}} (\frac{2\gamma}{r}), \tag{5.8}$$

[2]We are assuming the surface tension at the vapor-liquid-tube junction is approximately the same as the surface tension used in Sec. (5.1).

where in the last term we used $\rho_{vap} << \rho_{lig}$. Defining the pressure difference between inside and outside of the liquid droplet as $\delta P_{liq} = P_1 - P_2$, and using Eq. (5.3), we have

$$\delta P_{vap} = (\frac{\rho_{vap}}{\rho_{liq}})\delta P_{liq}. \tag{5.9}$$

We saw from Eq. (5.4) that δP_{liq} is generally quite small. Eq. (5.9) shows that δP_{vap}, while nonzero, is still smaller by the ratio ρ_{vap}/ρ_{liq}. Fermi puts this quantitatively by calculating the ratio $\delta P_{vap}/P_{vap}$ at $T = 300\,\mathrm{K}$. He gives

$$\frac{\delta P_{vap}}{P_{vap}} \approx \frac{10^{-7}}{r(\mathrm{cm})}. \tag{5.10}$$

For $r = 1\,\mathrm{mm}$, Eq. (5.10) says that δP_{vap} is a million times smaller than P_{vap} itself.

5.3 Radius and Terminal Velocity of Falling Raindrops

Raindrops typically form at a height that may vary up to several kilometers. They do so mainly by condensation of water vapor around a dust particle. The drops vary as they fall to Earth and, depending on their size, different methods are used to calculate the resistance they encounter in their descent. The drops may also undergo change of shape, but we follow Fermi in making the approximation that they retain their spherical form. He considers two limiting cases, one corresponding to very small raindrops and the other to larger ones as illustrations of two different mechanisms that act to determine the speed with which drops fall.

The first case is that of motion of a sphere of radius r moving at constant velocity through an infinite incompressible fluid of constant density and viscosity, a standard problem in transport phenomena (Huang 1987). The equation that describes such a state, the balance between the retarding force of the medium and the force of gravity, is known as Stokes's Law. It says

$$mg = 6\pi\nu r v, \tag{5.11}$$

where m is the mass of the drop, r is its radius, v, its terminal velocity, and ν is the coefficient of viscosity. Taking $m = \rho(4\pi r^3/3)$, and working in cgs units with $\rho = 1\,\mathrm{g/cm^3}$, we find the expression for the drop's terminal velocity to be

$$v(\mathrm{cm/s}) = \frac{2r^2\rho g}{9\nu} = 1.26 \times 10^6 (r(\mathrm{cm}))^2, \tag{5.12}$$

where we have employed the value of viscosity coefficient for air that Fermi gives: $\nu = 17.3 \times 10^{-5}\,\mathrm{g/s \cdot cm}$. This relation holds, however, only for the very smallest of drops because the viscosity retarding force given in Eq. (5.11) corresponds to just the first term in an expansion:

$$F = 6\pi\nu r v(1 + \frac{3\rho v r}{8\nu} + ...). \tag{5.13}$$

Its validity depends on the relative smallness of $\rho v r/\nu$, a quantity known as the Reynolds number (Landau and Lifshitz 1959). Since the viscosity of air is such a relatively small number, Stokes's law is only a good approximation for the case of for the smallest of raindrops, ones that are essentially mist-like, $r \leq 2.5 \times 10^{-3}\,\mathrm{cm}$. In this case the terminal velocity is small; a few cm/s (Van Boxel 1998).

For larger drops, terminal velocity is reached differently. The force of gravity is now balanced by an upward-directed aerodynamic drag. The equation describing the balance is

$$mg = \frac{4\pi r^3}{3}\rho g = \kappa\pi r^2 v^2, \tag{5.14}$$

where κ is the coefficient of air resistance. This is normally written as

$$\kappa = \frac{\rho_{air} C}{2}, \tag{5.15}$$

where C is a dimensionless quantity that is of order unity for air. We thus have

$$v = \sqrt{\frac{8gr}{3\rho_{air} C}} \approx 1500\sqrt{r} \cdot \mathrm{cm/s} \tag{5.16}$$

which gives $v \approx 7$ m/s for a raindrop with a 2 mm radius, a figure in rough agreement with experimentally observed values. It may also be worth noting that this formula, as expected, also accounts for the terminal velocity of hailstones.

Fermi quotes two values of terminal raindrop velocity, one at sea level and the other at an altitude of 4 km. He gives the first as $v = 1344\sqrt{r}$ and the second as $v = 2200\sqrt{r}$ at 4 km above sea level.The increase, 2200/1344, corresponds roughly to the decrease in the ratio of the square roots of ρ_{air} at the two altitudes, a dependence indicated in formula Eq. (5.16).

(8)

Coriolis forces on wind

Acceleration = centripetal acceleration + relative acceleration $+ 2\omega\wedge V_{rel}$

F + centrif force $= m A_{rel} + 2m\,\omega \wedge V_{rel}$

Project on horizontal plane $2\omega V \sin$ latitude

Order of magnitude of forces $\omega \sim \frac{2\pi}{T} = 7.3\times 10^{-5}$

1 cm Hg on 100 Km

Force from grad p wind bends towards the right in N hemisphere, " " left in S hemisphere

$-\frac{1}{\rho}$ grad p per unit mass

$$-\frac{1}{\rho}\,\text{grad}\,p = 2\omega v \sin\lambda$$

$$v = \frac{\text{grad}\,p}{2\omega\rho}$$

wind

Coriolis effects

CHAPTER 6

Coriolis Effects in the Earth's Atmosphere

6.1 The Coriolis Force

In this section, we give a brief review of the Coriolis force. This force plays an important role in understanding the physics of storms and hurricanes in the Earth's atmosphere. It is a "fictitious force," arising from the use of an accelerated or noninertial coordinate system. Such forces share with gravity the property of being proportional to the mass of the particle being considered, or the mass density in the dynamics of a fluid. Despite the nomenclature, the effects of the Coriolis force are very real.

We consider two coordinate systems; an inertial or space-fixed coordinate system with unit vectors $\hat{\boldsymbol{x}}, \hat{\boldsymbol{y}}, \hat{\boldsymbol{z}}$, and a rotating or body-fixed coordinate system with unit vectors $\hat{\boldsymbol{x}}', \hat{\boldsymbol{y}}', \hat{\boldsymbol{z}}'$. The body-fixed system is rotating with respect to the space-fixed coordinate system around the common z axis. The angular velocity, assumed constant, is ω. The unit vectors in the two systems are related by

$$\begin{aligned} \hat{\boldsymbol{x}}' &= \cos\phi\,\hat{\boldsymbol{x}} + \sin\phi\,\hat{\boldsymbol{y}}, \\ \hat{\boldsymbol{y}}' &= -\sin\phi\,\hat{\boldsymbol{x}} + \cos\phi\,\hat{\boldsymbol{y}}, \\ \hat{\boldsymbol{z}}' &= \hat{\boldsymbol{z}}, \end{aligned} \tag{6.1}$$

where $\phi = \omega t$.

A generic vector such as $\boldsymbol{r}$, the coordinate of a particle, can be expanded in either coordinate system;

$$\boldsymbol{r} = x\,\hat{\boldsymbol{x}} + y\,\hat{\boldsymbol{y}} + z\,\hat{\boldsymbol{z}} = x'\,\hat{\boldsymbol{x}}' + y'\,\hat{\boldsymbol{y}}' + z\,\hat{\boldsymbol{z}} = \boldsymbol{r}', \tag{6.2}$$

where, since the z axis is common, there is no need for primes on z components or unit vectors. Furthermore, note that $\boldsymbol{r}$ and $\boldsymbol{r}'$ are just different notations for the same vector.

Subtleties arise when computing time derivatives of vectors. It is necessary to distinguish the time rate of change with respect to the space-fixed system from one taken with respect to the body-fixed system. When $\boldsymbol{r}$ is differentiated in the space-fixed system, the result is

$$\frac{d}{dt}\boldsymbol{r} = \frac{dx}{dt}\hat{\boldsymbol{x}} + \frac{dy}{dt}\hat{\boldsymbol{y}} + \frac{dz}{dt}\hat{\boldsymbol{z}}. \tag{6.3}$$

There is no question of taking time derivatives of $\hat{\boldsymbol{x}}, \hat{\boldsymbol{y}}, \hat{\boldsymbol{z}}$; they are unit vectors in an inertial system and therefore constant. Now consider time derivatives from the viewpoint of a body-fixed observer. Such an observer will naturally regard the body-fixed unit vectors as constant, and define his rate of change of $\boldsymbol{r}$ as

$$\boldsymbol{v}' \equiv \frac{d\,'}{dt}\boldsymbol{r} = \frac{dx'}{dt}\hat{\boldsymbol{x}}' + \frac{dy'}{dt}\hat{\boldsymbol{y}}' + \frac{dz}{dt}\hat{\boldsymbol{z}}. \tag{6.4}$$

The difference between the rates of change from Eqs. (6.3) and (6.4) comes about because the body-fixed unit vectors $\hat{\boldsymbol{x}}'$ and $\hat{\boldsymbol{y}}'$ are not constants in the space-fixed system. The rates of change of the body-fixed unit vectors from the viewpoint of the space-fixed system are

$$\frac{d}{dt}\hat{\boldsymbol{x}}' = \boldsymbol{\omega} \wedge \hat{\boldsymbol{x}}', \tag{6.5}$$

$$\frac{d}{dt}\hat{\boldsymbol{y}}' = \boldsymbol{\omega} \wedge \hat{\boldsymbol{y}}',$$

where the angular velocity vector is $\boldsymbol{\omega} = \omega\hat{\boldsymbol{z}}$. With the aid of Eq. (6.5), it is easy to relate the two rates of change of $\boldsymbol{r}$. We have

$$\frac{d}{dt}\boldsymbol{r} = \frac{d}{dt}'\boldsymbol{r} + \boldsymbol{\omega} \wedge \boldsymbol{r}, \tag{6.6}$$

or

$$\boldsymbol{v} = \boldsymbol{v}' + \boldsymbol{\omega} \wedge \boldsymbol{r}. \tag{6.7}$$

Symbolically, the two rates of change are related by

$$\frac{d}{dt} = \frac{d}{dt}' + \boldsymbol{\omega} \wedge . \tag{6.8}$$

In the space-fixed frame, Newton's law takes the simple form,

$$\boldsymbol{F} = m\frac{d^2\boldsymbol{r}}{dt^2} = m\frac{d}{dt}(\frac{d}{dt}\boldsymbol{r}). \tag{6.9}$$

Substituting from Eq. (6.6) gives

$$\boldsymbol{F} = m\frac{d}{dt}(\frac{d}{dt}'\boldsymbol{r} + \boldsymbol{\omega} \wedge \boldsymbol{r}). \tag{6.10}$$

Applying Eq. (6.8), we have

$$\boldsymbol{F} = m(\frac{d}{dt}' + \boldsymbol{\omega}\wedge)(\frac{d}{dt}'\boldsymbol{r} + \boldsymbol{\omega} \wedge \boldsymbol{r}). \tag{6.11}$$

Expanding the right-hand side gives

$$\boldsymbol{F} = m\left\{\frac{d'}{dt}(\frac{d'}{dt}\boldsymbol{r}) + 2\boldsymbol{\omega}\wedge\frac{d'}{dt}\boldsymbol{r} + \boldsymbol{\omega}\wedge(\boldsymbol{\omega}\wedge\boldsymbol{r})\right\}. \tag{6.12}$$

The first term on the right side of Eq. (6.12) comes from the acceleration of the particle as seen in the body-fixed frame, the middle term is the Coriolis force, and the last term is the centrifugal force. To write this equation in a more conventional way, define the body-fixed acceleration by

$$\boldsymbol{a}' = \frac{d'}{dt}(\frac{d'}{dt}\boldsymbol{r}) = \frac{d'}{dt}\boldsymbol{v}'. \tag{6.13}$$

Then using $\boldsymbol{r} = \boldsymbol{r}'$, we have

$$\boldsymbol{F} = m\boldsymbol{a}' + 2m\boldsymbol{\omega}\wedge\boldsymbol{v}' + m\boldsymbol{\omega}\wedge(\omega\wedge\boldsymbol{r}'). \tag{6.14}$$

Defining

$$\boldsymbol{F}_{eff} = \boldsymbol{F} - 2m\boldsymbol{\omega}\wedge\boldsymbol{v}' - m\boldsymbol{\omega}\wedge(\omega\wedge\boldsymbol{r}'), \tag{6.15}$$

Newton's law in the body-fixed coordinate system takes the simple form (Goldstein 1999)

$$\boldsymbol{F}_{eff} = m\boldsymbol{a}'. \tag{6.16}$$

For the important case of the rotating Earth, expressing $\boldsymbol{r}$ in terms of components parallel to and perpendicular to $\boldsymbol{\omega}$, we have

$$\boldsymbol{r} = r_{\|}\hat{\boldsymbol{\omega}} + \boldsymbol{r}_{\perp}. \tag{6.17}$$

Using this form for $\boldsymbol{r}$, the last term in Eq. (6.15) becomes

$$-m\boldsymbol{\omega}\wedge(\omega\wedge\boldsymbol{r}) = m\omega^2\boldsymbol{r}_{\perp}, \tag{6.18}$$

which is just the centrifugal force as given in Eq. (2.12).

6.2 Coriolis Effects in the Earth's Atmosphere

All equations in this section are in a rotating coordinate system rigidly attached to the Earth, so primes on vectors and time derivatives are dropped. For a moving element of fluid, instead of a mass m subject to a force $\boldsymbol{F}$, we consider the force per unit volume $\boldsymbol{f}$[1] acting on the mass density ρ. The analog of Newton's second law is Euler's fluid equation, which in our rotating coordinate system takes the form

$$\rho(\frac{\partial \boldsymbol{v}}{\partial t} + \boldsymbol{v} \cdot \boldsymbol{\nabla} \boldsymbol{v}) = \boldsymbol{f} - 2\rho\boldsymbol{\omega} \wedge \boldsymbol{v} - \rho\boldsymbol{\omega} \wedge (\boldsymbol{\omega} \wedge \boldsymbol{r}). \qquad (6.19)$$

In Eq. (6.19),

$$\frac{\partial \boldsymbol{v}}{\partial t} + \boldsymbol{v} \cdot \boldsymbol{\nabla} \boldsymbol{v}$$

is the local acceleration moving with the fluid, and the right-hand side of the equation is the analog for a fluid of Eq. (6.15).

Treating the atmosphere as an ideal fluid (ignoring viscosity), the force/volume is

$$\boldsymbol{f} = -\boldsymbol{\nabla} p - \rho \boldsymbol{\nabla} V_g, \qquad (6.20)$$

where p is the pressure, and V_g is the gravitational potential due to the masses in the Earth. As in Sec. (2.2), we combine the gradient of V_g and the centrifugal force term, defining

$$\boldsymbol{g} \equiv -\boldsymbol{\nabla} V_g - \boldsymbol{\omega} \wedge (\boldsymbol{\omega} \wedge \boldsymbol{r}). \qquad (6.21)$$

With these definitions, Eq. (6.19) reads

$$\rho(\frac{\partial \boldsymbol{v}}{\partial t} + \boldsymbol{v} \cdot \boldsymbol{\nabla} \boldsymbol{v}) = -\boldsymbol{\nabla} p - 2\rho\boldsymbol{\omega} \wedge \boldsymbol{v} + \rho \boldsymbol{g}. \qquad (6.22)$$

The middle term on the right-hand side of Eq. (6.22) is the Coriolis force/volume. It plays a crucial role in many atmospheric phenomena.

[1]The force/mass $\boldsymbol{f}$ is not to be confused with the Coriolis parameter f, defined below.

Wind Circulation around Low- and High-Pressure Regions This section deals with the direction that wind circulates around points of high and low pressure.[2] It depends on whether the center of the storm is a local maximum or minimum of pressure, and whether the storm is in the upper or lower hemisphere.

Storms typically form a cylindrical structure, the axis of which is aligned with the local vertical direction. Their centers do of course move, but very slowly compared to the wind speeds in the storm, so we assume the storms are stationary, with their axes along the local vertical direction.

For describing the coordinates of a storm, it is convenient to take the z axis of the Earth-fixed coordinate system pointing in the local vertical direction, parallel to the axis of the storm. Although storms can extend over hundreds of kilometers, the extent of even a very large storm is still very small compared to the radius of the Earth. The curvature of the Earth therefore plays no role in describing the storm, and $\boldsymbol{g}$ in Eq. (6.22) can be taken to be a constant vector pointing inward along the z axis.

A point in the storm is located by giving z, the vertical height above the surface of the Earth, and $\boldsymbol{r}$, the horizontal vector from the storm axis to the point of interest.[3] A section of the storm at a definite elevation z forms a horizontal plane perpendicular to the z axis. The plane tangent to the Earth at $z = 0$ is known as the "f plane" in meteorology. For an axially symmetric storm, cylindrical coordinates are very natural, the coordinates being $z, r = |\boldsymbol{r}|$, and λ, where λ is an angle specifying the angle $\boldsymbol{r}$ makes with an axis in the same horizontal plane. The "radial direction" is simply the

---

[2]The storms considered in this section are variously known as "tropical cyclones," "cyclonic storms," "cyclones," and "hurricanes."

[3]Note that $\boldsymbol{r}$ is *not* the usual radial vector in spherical coordinates.

direction of $\boldsymbol{r}$. Fig. (6.1) shows schematically a cyclonic storm, the f plane, and the location of a parcel of atmosphere inside the storm.

Writing the velocity of a point in the storm in cylindrical coordinates, we have

$$\boldsymbol{v} = u\hat{\boldsymbol{r}} + v\hat{\boldsymbol{\lambda}} + w\hat{\boldsymbol{z}}. \tag{6.23}$$

In a cyclonic storm, the radial velocity u is the velocity of the wind heading toward or away from the storm axis, the azimuthal velocity v is the velocity of the wind circulating around the storm axis, and w is the wind velocity in the vertical direction, traveling up or down in the axial direction.

The Coriolis term in Eq. (6.22) is small compared to $|\boldsymbol{g}|$. Suppose a hurricane has wind speeds of 100 mph, or 44 m/s. The duration of a day is $T_d = 86,400\,\text{s}$, so the Coriolis term is maximized by

$$|\boldsymbol{\omega} \wedge \boldsymbol{v}| \leq \frac{2\pi}{T_d}|\boldsymbol{v}| = 3.2 \times 10^{-3}\,\text{m/s}^2, \tag{6.24}$$

which is completely negligible compared to $|\boldsymbol{g}| = 9.8\,\text{m/s}^2$. As a result, Coriolis forces in the vertical direction can be ignored. However, Coriolis forces in the horizontal direction play an important role in hurricane dynamics. Such terms could in principle come from the horizontal component of $\boldsymbol{\omega}$ crossed into the vertical term in $\boldsymbol{v}$. However, the vertical velocity w is the smallest wind velocity in a cyclic storm, typically $\leq 1\,\text{m/s}$, so this term can also be neglected. The horizontal Coriolis terms of interest are generated by replacing $\boldsymbol{\omega}$ with its vertical component, $\boldsymbol{\omega} \wedge \boldsymbol{v} \to \omega \cos\theta(\hat{\boldsymbol{z}} \wedge \boldsymbol{v})$. In meteorology, what is known as the "Coriolis parameter" is defined by $f \equiv 2\omega\cos\theta$, where $\omega = 2\pi/T_d$ is the angular velocity of the Earth and θ is the polar angle as defined above. Note that due to the factor of $\cos\theta$ in the definition of

f, f is positive in the Northern Hemisphere, negative in the Southern Hemisphere, and zero on the equator. In terms of f, the Coriolis term on the right side of Eq. (6.22) becomes a combination of radial and azimuthal terms,

$$-2\rho\omega\cos\theta\hat{\boldsymbol{z}}\wedge\boldsymbol{v} = \rho f(v\hat{\boldsymbol{r}} - u\hat{\boldsymbol{\lambda}}). \tag{6.25}$$

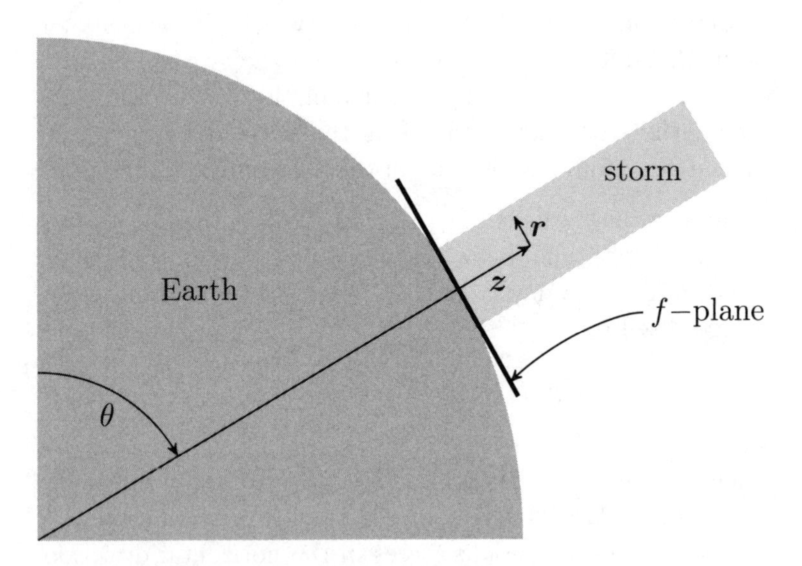

Figure 6.1: Coordinates for a cyclonic storm

The dominant feature of a cyclonic storm is, as the name implies, rotation around the center of the storm. The largest component of $\boldsymbol{v}$ is v. As mentioned above, the vertical velocity w is quite small, but it is also true that the radial velocity u is much smaller than the azimuthal velocity v. As a result, insight into the structure of a storm can be gained by setting $u = w = 0$ and looking for a steady-state solution. The solution of Eq. (6.22) obtained by doing so is termed "gradient wind balance." The vertical component of Eq. (6.22) reduces to the barometric equation

$$-\frac{\partial p}{\partial z} = \rho g. \tag{6.26}$$

The radial equation is

$$\frac{v^2}{r} + fv = \frac{1}{\rho}\frac{\partial p}{\partial r}. \tag{6.27}$$

The v^2/r term in Eq. (6.27) is the centripetal acceleration familiar in elementary mechanics. It arises from the radial component of the $\boldsymbol{v} \cdot \boldsymbol{\nabla} \boldsymbol{v}$ term in Eq. (6.22),

$$\hat{\boldsymbol{r}} \cdot (\boldsymbol{v} \cdot \boldsymbol{\nabla} \boldsymbol{v}) = \hat{\boldsymbol{r}} \cdot (v \frac{1}{r}\frac{\partial}{\partial \lambda} v \hat{\boldsymbol{\lambda}}) = -\frac{v^2}{r}. \tag{6.28}$$

Let us consider the sizes of the terms in Eq. (6.27) for a typical tropical cyclone. Taking $\theta = 45°$ for definiteness, the Coriolis parameter is $f \sim 10^{-4}\,\mathrm{s}^{-1}$. At an intermediate distance from the axis of such a storm, the velocity might be as large as $v \sim 40\,\mathrm{m/s}$, and for $r \sim 40\,\mathrm{km}$, the centripetal term v^2/r is dominant, approximately ten times larger than the Coriolis term. However, as the radius increases, the velocity v decreases, and at $r \sim 200\,\mathrm{km}$, $v \sim 10\,\mathrm{m/s}$, the Coriolis term is twice as large as the centripetal term. At still larger values of r, the Coriolis term dominates. At very large values of r, where the Coriolis term dominates, Eq. (6.27) becomes

$$fv \sim \frac{1}{\rho}\frac{\partial p}{\partial r}. \tag{6.29}$$

We see that the sign of fv is correlated with the sign of $\partial p/\partial r$. For a storm which surrounds a pressure minimum, we expect that $\partial p/\partial r > 0$, so fv must be positive. In the Northern Hemisphere, where $f > 0$, v must be positive as well; the sense of rotation around the pressure minimum is counterclockwise. In the Southern Hemisphere, where $f < 0$, v must be negative; the sense of rotation around a pressure minimum is clockwise. Both of these behaviors have been extensively documented by satellite photos. Pressure maxima are more rare, but do occur, e.g., in Siberia. For a storm surrounding a pressure maximum, the sense of rotation is clockwise in the Northern Hemisphere, counterclockwise in the Southern.

It is of interest to estimate the kinetic energy of rotation in a cyclonic storm. At intermediate distances from the axis of the storm, the whole system is basically in solid body rotation. Taking the storm to be a cylinder of height $h \sim 15\,\text{km}$, radius $\sim 30\,\text{km}$, regarding the system as a giant cylinder rotating at a frequency $\sim f$, the kinetic energy of rotation is roughly $6 \times 10^{13}\,\text{J}$, quite comparable to the energy released in the Hiroshima atomic bomb!

Jet Streams The temperature of the Earth's atmosphere is higher near the equator than near the North or South Pole and the resulting lack of thermal equilibrium causes a flow of air toward the poles in both hemispheres. The Coriolis force then deflects these poleward flows. To get the overall direction of the resulting force, let us assume the flow of air is strictly toward the poles in both hemispheres. This can be written as

$$\boldsymbol{v} = v_\theta \hat{\boldsymbol{\theta}}, \tag{6.30}$$

where $\hat{\boldsymbol{\theta}}$ and $\hat{\boldsymbol{\phi}}$ are the usual unit vectors for polar and azimuthal angles. In the Northern Hemisphere, poleward flow implies $v_\theta < 0$; the flow is from values of $\theta < 90°$ toward $\theta = 0$. In the Southern Hemisphere, flow toward the South Pole corresponds to $v_\theta > 0$; the flow is from an angle $\theta > 90°$ toward $\theta = 180°$. So v_θ has opposite sign in the two hemispheres.

Taking account of the sign change of $\cos\theta$ in the two hemispheres, we see from Eq. (6.22) that the resulting Coriolis force is along the $\hat{\phi}$ or eastern direction in *both* hemispheres; the Coriolis force on the poleward flow is always toward the east, regardless of hemisphere. The resulting deflection gives the general direction of the jet streams in both Northern and Southern Hemispheres. While this argument does give the correct west-to-east direction of the jet streams, there are additional complex behaviors. Perhaps the most important of these is that jet streams, in addition to their general eastward flow, have oscillations in latitude. These have a major effect on the weather in the northern United States. When the jet stream "dips" into the northern US, bitter cold winter weather is the usual result. The full physics of the jet streams was first elucidated by the Swedish-American meteorologist Carl Gustav Rossby in 1939 (Rossby 1939), and the term "Rossby wave" is associated with the oscillations of the jet stream in the atmosphere. Rossby waves also occur in the ocean.

6.3 Hurricanes

Hurricane as an Example of a Carnot Cycle Hurricanes illustrate how a simple construct can be useful in organizing and describing the features of a complex physical phenomenon. In this case the construct is the one discussed in Sec. (3.3), the Carnot cycle.

Hurricane, a term used in the North Atlantic for a storm with average winds over 30m/s, is a weather system generated in tropical zones by a low-pressure area rising over the ocean being acted upon by the Coriolis force that converts the inward rush of air toward the eye of the storm into a circulating motion. In the western Pacific such systems are known as *typhoons:* the general term is *severe tropical cyclone.* The characteristics are the same everywhere.

In applying the steps of a Carnot cycle to such a storm, the ocean acts as the heat source of an isothermal step at a constant temperature $T_2 \approx 300\,\mathrm{K}$, the heat energy is provided by the ocean through the evaporation of water (Emanuel 1991). This is followed in step two of the Carnot cycle by the rapid adiabatic expansion of the air as it rises to an altitude of order 15 km, a height where the ambient temperature is $T_1 \approx 200\,\mathrm{K}$. In step three the air is cooled by radiation to outer space and compressed; in step four the air descends adiabatically to sea level and the cycle then resumes.

Calculation of Wind Velocity As an example of the analogy's usefulness, we will use it to obtain a formula for the wind velocity in terms of the air's heat gains and loses. The work done in this cycle goes into the frictional dissipation that occurs as the air blows over the ocean surface. Of course frictional dissipation is also the cause of the destruction wreaked by the storm once it makes landfall. Since the wind's drag force $F_{drag} \propto v^2$, work is done by the wind over the ocean at a rate of

$$dW/dt = -F_{drag}v = -Av^2 \cdot v,$$

where A is a constant. Heat is given off to the wind by the ocean at a rate Bv, with B also a constant, representing the lack of equilibrium between the ocean and the wind. This heat can optimally be converted into work during the cycle at the rate proportional to η, the Carnot cycle's thermal efficiency,

$$\eta = \frac{T_2 - T_1}{T_2}.$$

At equilibrium we may equate the rate at which heat is converted into work with the rate at which work is done through frictional dissipation. This implies that

$$Av^3 = \eta B v.$$

We can therefore expect a hurricane velocity of

$$v^2 = \eta(B/A) \implies v = \sqrt{\frac{B\eta}{A}} = \sqrt{\frac{T_2 - T_1}{T_1}(B/A)}.$$

If, as we said earlier, $T_2 \approx 300\,\mathrm{K}, T_1 \approx 200\,\mathrm{K}$, we find

$$v \approx \sqrt{B/2A},$$

in approximate agreement with experimental observation.

Surface of the Earth (9)

5.10×10^{18} cm^2

29% dry land

71% sea

Hypsographic curve

average level of crust −2440 m

average depth of sea −3800 m

100 200 300 400 500 M km^2

One liter sea water contains 35 g. of various salts out of which about 27 g are NaCl

Thermal balance of sea

	g cal/cm^2 day		
Sun radiation	300	Radiation outside	660
rad. from the atmosph	500	Convection to air	40
Internal heat	(.2)	Evaporation	100
	800	(1 mm = 60) (60 cm/year ~ 100 cal)	800

insert at beginning

Temperature equilibrium of the Earth surface

From Sun 2 cal/min

$\sigma = 5.68 \times 10^{-5}$ erg/cm^2 deg^4 Stefan constant

$4\pi R^2 \sigma T^4 = \pi R^2 \times$ energy/cm^2 sec

$T^4 = 61.7 \times 10^8$ $T = 280\,°K = 7°C = 45°F$

Thermal properties of Earth's surface above and below sea level

CHAPTER 7

Thermal Properties and Radiation at the Earth's Surface

7.1 Surface of the Earth

The surface of the solid Earth lies partly below sea level and partly above sea level. As a glance at a desktop globe makes clear, the greater fraction of the Earth's crust lies below sea level. In terms of area, roughly 29% of the surface area is above sea level and the remaining 71% is below sea level. The "hypsographic curve" is shown in Fig. (7.1). This illustrates what percentage of the Earth's solid surface lies above a given elevation. The percentage starts from zero at the top of Mt. Everest, at an elevation of 8.848 km, and reaches 100% at the bottom of the Mariana Trench in the Pacific Ocean, at a depth or negative elevation of 10.984 km. From the top of Mt. Everest to the bottom of the Mariana Trench spans a

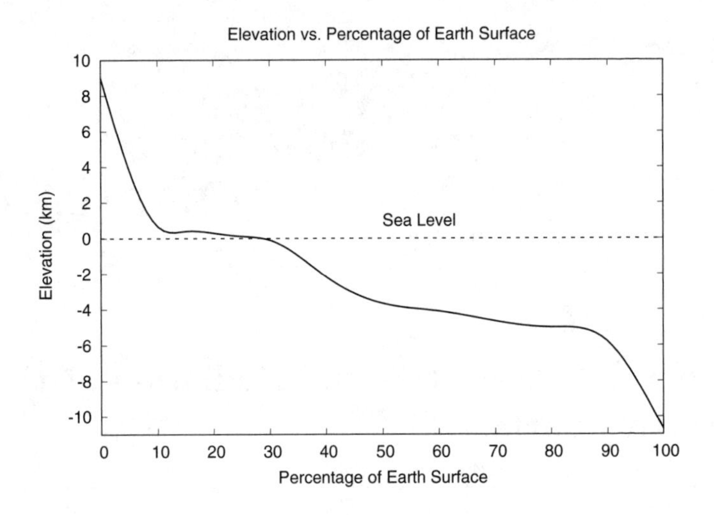

Figure 7.1: Hypsographic curve-
Source: NOAA

distance of roughly 20 km, a large distance on a human scale, but less than 1% of the Earth's radius, so on a planetary scale, the Earth's surface is relatively smooth.

7.2 Crust of the Earth

The crust is a layer of rock lying above, or one might even say "floating" on, the Earth's mantle, a layer that near the top can behave like a very viscous fluid. The oceanic crust is 5–10 km thick with an average density of $3.0\,\mathrm{g/cm^3}$ while the continental crust is typically less dense, with an average density of $2.7\,\mathrm{g/cm^3}$ and a thickness that ranges from 30 to 50 km. This is a dynamic system with parts of the crust sinking deeper into the mantle until a condition of equilibrium known as isostasy is reached. In Sec. (12.10) we shall go into more detail about the behavior of the continental crust.

7.3 Composition of the Oceans

Each liter of seawater contains approximately 35 g of dissolved salts. The average density of seawater is $1.027\,\mathrm{g/cm^3}$. The main components of the salts, 27 out of the 35 g, are sodium and chlorine ions. Sulfur, magnesium, calcium, and potassium ions constite almost all of the remainder. In Chap. (8), we will discuss the thermal properties of ocean water.

7.4 Temperature Equilibrium of the Earth's Surface

Solar radiation is the main source of the Earth's energy. The total power emitted by the Sun in the form of electromagnetic waves is called the *solar luminosity*, $Q_\odot$. It is determined from the average power/area reaching the Earth. This quantity is known as the *solar constant*, $S_\odot$. Attempts to determine the solar constant began in the late nineteenth century. At present, satellite measurements measure both the time dependence and the mean value of $S_\odot$ with great accuracy. When sunspots are at a minimum in the solar cycle, the mean value of the solar constant is

$$S_\odot = 1361\,\mathrm{W/m^2}. \qquad (7.1)$$

This is roughly the same power that would result from thirteen one-hundred-watt light bulbs held directly over an area of one square meter.

To evaluate $Q_\odot$, we use the fact that the total power emitted by the Sun passes through every sphere surrounding the Sun. For a sphere of radius r, the resulting intensity is

$$(\frac{Q_\odot}{4\pi r^2})\mathrm{W/m^2}. \tag{7.2}$$

Taking r to be the Earth-Sun distance $R_{es} = 149.6 \times 10^6$ km, $Q_\odot$ is

$$Q_\odot = 4\pi(R_{es})^2 S_\odot = 3.828 \times 10^{26}\,\mathrm{W}. \tag{7.3}$$

The solar luminosity is constant to within 0.1% over the last several 11-year solar sunspot cycles.

7.5 Black Bodies and the Temperature of Sun and Earth

A so-called black body is simply one which totally absorbs every wavelength of radiation incident upon it. During the nineteenth century, physicists proved a number of general theorems about black bodies in thermodynamic equilibrium, a state in which the black body emits the same amount of radiation that it absorbs. It should be noted that the term "black body" is a bit of a misnomer. A totally absorbing body will only appear to be black when its temperature is relatively low, such as room temperature. The body will begin to glow when its temperature is high enough so that a substantial amount of energy is emitted in the visible part of the spectrum.

The intensity I (units: power/area/s) radiated from a black body is known as the Stefan-Boltzmann law, which states

$$I = \sigma T^4, \tag{7.4}$$

where σ is the Stefan-Boltzmann constant. The value of σ is

$$\sigma = 5.67 \times 10^{-8}\mathrm{W/m^2/\,(K)^4}.$$

The value of σ, known empirically in the nineteenth century, had been shown by general thermodynamic arguments to be independent of the substance forming the black body. Early in the twentieth century, after the discovery of the quantum nature of light by Max Planck, it was realized that σ can be expressed in terms of the velocity of light, Boltzmann's constant, and Planck's constant. See Eq. (7.14).

Assuming the Sun and Earth are black bodies, their temperatures can be estimated using the Stefan-Boltzmann law. The Sun's radius is $R_s = 695,700\,\text{km}$. Dividing $Q_\odot$ by the Sun's area, we have an expression for the intensity emitted by the Sun. Applying the Stefan-Boltzmann law, we have

$$I_s = \frac{Q_\odot}{4\pi R_s^2} = \sigma(T_s)^4. \tag{7.5}$$

Solving for T_s gives

$$T_s = 5772\,K. \tag{7.6}$$

Another law, derived from thermodynamic principles toward the end of the nineteenth century, provides an independent estimate of the temperature of the solar surface. Wien's displacement law (Landau and Lifshitz 1980) gives a relation between the wavelength of the body's maximum intensity of radiation and its temperature in kelvins.

$$\lambda_{max} \cdot T = 2.898 \times 10^7\,\text{Å}\,\text{K}, \tag{7.7}$$

where Å stands for the angstrom unit of 10^{-8} cm. The wavelength range of visible light is 3900 Å to 7000 Å, the midpoint being 5450 Å. From Wien's law, a black body with $\lambda_{max} = 5450$ Å would have $T = 5317\,\text{K}$, a value within 10% of Eq. (7.6).

Assuming also that the Earth is a black body, and that the incident intensity arises solely from incoming solar energy, the total incident solar energy is the product of the solar constant and πR_e^2, the area the Earth presents to the incoming radiation. If the Earth were to re-radiate this energy as a black body, we would have

$$S_\odot \pi R_e^2 = 4\pi R_e^2 \sigma T_e^4. \tag{7.8}$$

Note that in the left side of Eq. (7.8), it is the area of the Earth's cross-section that enters, while on the right-hand side it is the total surface area of the Earth. Solving this equation for T_e gives

$$T_e = 278\,\mathrm{K}, \tag{7.9}$$

again a reasonable value. This calculation ignores the presence of the Earth's atmosphere, which reflects part of the Sun's energy. The fraction, known as the *albedo* and denoted as α, has a value of $\alpha = 0.3$ for the Earth. The atmosphere itself can be modeled as a so-called *grey body*, which satisfies a modified form of the Stefan-Boltzmann law,

$$I_{grey} = \epsilon \sigma T^4, \tag{7.10}$$

where ϵ is the emissivity of the atmosphere, for which a reasonable value is $\epsilon = 0.8$. It is also worth noting, as Fermi points out, that if the absorption of solar radiation were reduced by a factor $A \leq 1$ and the Earth's radiation by a factor $a \leq 1$, Eq. (7.8) should be modified by multiplying the left side by A and the right side by a. The result would be

$$T_e = 278\,K \cdot (A/a)^{1/4}. \tag{7.11}$$

As a rough estimate of these effects, set $A = 1 - \alpha = 0.7$, and $a = \epsilon = 0.8$. The effect is to reduce the temperature of the Earth's atmosphere from $T_e = 278\,\mathrm{K}$ to $T_e = 268\,\mathrm{K}$.

The assumption that the Earth is effectively a black body is only a rough approximation. In particular the structure of many molecules makes for a very selective absorption. An example is water, which absorbs strongly in the infrared and is relatively transparent in the visible range. Despite these complications, assuming black body behavior for both Earth and Sun gives a reasonable estimate for the mean temperature of the Earth's atmosphere.

With regard to the actual balance between absorption of solar radiation and re-radiation by the Earth, a full understanding is complicated by strong dependence on cloud cover, the mix of gases in the atmosphere, and a myriad of other factors. A schematic picture of how that balance is achieved is presented in Fig. (7.2).

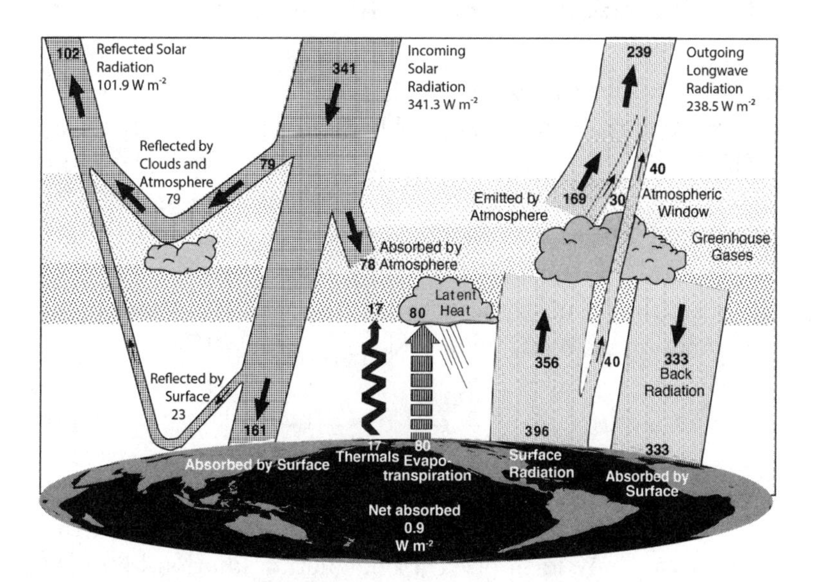

Figure 7.2: Energy balance; figure courtesy of Kevin Trenberth, John Fasullo, and Jeff Kiehl, USNCAR (National Center for Atmospheric Research)

Fermi provides a few lines of estimates of thermal balance,

displaying how intake and outtake roughly match for the case of the sea. He also notes a striking fact that we will return to and discuss in detail in Chap. (15), namely that the heat reaching the surface from the Earth's interior, roughly due in equal parts to heat from the Earth's formation and decay of radioactive isotopes, is negligible compared to heat from the Sun, less than one part in a thousand. It is nevertheless crucial because it controls the geological features of the Earth: plate tectonics, igneous activity, etc. Rock is, however, such a poor conductor that the Earth's surface is practically insulated from it.

7.6 Planck's Spectral Radiance Formula

In 1900 Max Planck wrote his famous formula, marking the beginning of the quantum era. The formula is for the spectral radiance of a black body at temperature T. The term "radiance" implies power per unit area per unit solid angle. The addition of "spectral" means per unit frequency. The SI unit for spectral radiance is watts/meter^2/steradian/frequency. Denoting the spectral radiance by $B(\nu, T)$, Planck's formula as a function of frequency ν and temperature T is

$$B(\nu, T) = \frac{2h\nu^3}{c^2}[\exp(\frac{h\nu}{kT}) - 1]^{-1}, \qquad (7.12)$$

where k is Boltzmann's constant, c is the velocity of light, and h is Planck's constant. Planck's derivation required the concept that any form of electromagnetic radiation be made up of packets of energy, photons, each one having an energy $h\nu$.

Integrating $B(\nu, T)$ over all frequencies leaves a quantity with units of watts/meter^2/steradian. A subsequent solid angle integration would produce a quantity with the same units as occur in the Stefan-Boltzmann law. The actual relation between the Stefan-Boltzmann law and Planck's formula is

$$\sigma T^4 = \int_0^{2\pi} d\phi \int_0^{\pi/2} \sin\theta \cos\theta d\theta \int_0^{\infty} d\nu B(\nu, T). \qquad (7.13)$$

The angular integral in Eq. (7.13) has two features worth noting: (1) the upper limit of the θ integral is $\pi/2$ rather than π; and (2) there is a factor of $\cos\theta$ in the integrand. Both can be understood by considering a cavity containing black body radiation, with a small opening of area ΔA, from which black body radiation escapes. Erecting an outward normal to ΔA, it is clear that radiation which escapes the cavity must have polar angle θ in the range 0 to $\pi/2$. The factor of $\cos\theta$ can be understood by considering the current of photons headed from inside the cavity toward the opening ΔA. Let $e(\nu, T)$ be the energy density of photons/solid angle/frequency in the cavity. This is related to $B(\nu)$ by $e(\nu, T)c = B(\nu, T)$, where c is the velocity of light. The energy which arrives at ΔA per unit time/area/solid angle/frequency is $e(\nu, T)(c\cos\theta)$. It is evident that the photon energy current is proportional to the velocity with which the photons approach the opening. For a photon approaching the opening at angle θ to the normal, this velocity is $c\cos\theta$. This is the origin of the $\cos\theta$ in the integral of Eq. (7.13).

The integral in Eq. (7.13) is a standard one. The final result is

$$\sigma T^4 = \frac{2k^4\pi^5 T^4}{15c^2h^3}. \qquad (7.14)$$

Eq. (7.14) is a striking result in that it shows that constant σ can be evaluated via Planck's formula in terms of other known physical constants, including Planck's constant. Since Wien's displacement law is simply a statement of the location of the frequency peak of radiation from a black body, it must be equivalent to

$$\frac{\partial B(\nu,T)}{\partial \nu} = 0.$$

The constant in Wien's law is therefore also given by the fundamental parameters in Planck's law. It is worth noting that the Planck distribution for sunlight regarded as a function of wavelength has a broad maximum that closely matches the visible range of 3900 Å to 7000 Å.

7.7 Special Cases

In order to illustrate the range of the formulas derived in the previous section, Fermi shows how they could be used to consider the completely unphysical case in which absorption and emission of radiation from the Sun by the Earth only took place at a single frequency, ν_0. To do so we would have to replace Eq. (7.8) by absorption and emission at that single frequency,

$$B(\nu_0, T_s)\frac{4\pi R_s^2}{4\pi d^2}\pi R_e^2 = 4\pi R_e^2 B(\nu_0, T_e), \tag{7.15}$$

where d is equal to the distance between the Earth and the Sun (149.5×10^6 km). Rearranging this equation slightly, we have

$$4\frac{d^2}{R_s^2} = \frac{\exp(\frac{h\nu_0}{kT_e}) - 1)}{\exp(\frac{h\nu_0}{kT_s}) - 1)}. \tag{7.16}$$

The left-hand side is approximately equal to 18.6×10^4. Fermi looked at two limits, one in which $h\nu_0/kT_e \ll 1$ and the other in which $h\nu_0/kT_e \gg 1$. In the first case we see

$$18.6 \times 10^4 \approx \frac{T_s}{T_e} \Rightarrow T_e \approx 0.03\,\mathrm{K}. \tag{7.17}$$

Of course the Earth's temperature is so low because the photons being absorbed have very low energy. Conversely, for $h\nu_0/kT \gg 1$ the equality of Eq. (7.16) implies $T_e \approx T_s$. In one of these examples the Earth's temperature is almost absolute zero and in the other it lies close to that of the Sun.

Average vertical temperature

depth	temp
0	16°.0
100	15.8
200	15.5
400	13.7
600	9.9
800	5.1
1000	3.8
2000	3.1
3000	2.8
4000	2.5

heat leaking inside

$$-K_{eff}\frac{dt}{dz} = -10 \times \frac{1}{10000} = -10^{-3}\ \mathrm{cal/cm^2\,sec}$$

$$= 86\ \mathrm{cal/cm^2\,day}$$

hence lateral convection

Sinking of cold waters when t_{surf} is low ~~goes below~~ 4°

gives t_{deep} = 4 sweet, 2.5 sea water

Discussion of salt equilibrium

Tidal ~~potential~~

Mass of Moon 0.01227 Earth Sun 2×10^{33}

distance Moon 384000 km Sun ?

Ocean temperature as function of depth

CHAPTER 8

Thermal Properties of the Ocean

The oceans of the Earth form an enormously complex system whose behavior has important effects on a huge range of human activities. Oceans contain flows of both seawater and heat, e.g., the Gulf Stream. They are acted upon by solar energy, wind, and freshwater input. Physical quantities relevant for describing ocean behavior are temperature, fluid velocity, seawater density, and salinity.

Our treatment of the ocean's thermal behavior is necessarily brief. In the next two sections, Secs. (8.1) and (8.2), we introduce the notion of heat current and derive an equation that can describe the flow of heat in the ocean. In Sec. (8.3), we present Fermi's application of the heat equation to the diurnal behavior of the ocean near its surface. Finally, in Secs. (8.4) and (8.5), we give an overview of the variation of temperature with ocean depth and ocean circulation.

8.1 Heat Current and Thermal Conductivity

The direction of heat flow in a body is from higher to lower temperatures. The heat current density, $\boldsymbol{q}$, normally specified in units of watts per square meter ($\mathrm{W/m^2}$), measures the rate at which heat flows across unit area. In the simplest situation (no convection or mixing), the heat current is linear in the temperature gradient,

$$\boldsymbol{q} = -\kappa \boldsymbol{\nabla} T, \tag{8.1}$$

where κ is the heat conductivity of whatever material is conducting heat. The SI unit of κ is W/m K.[1] Examples of κ, all in W/m K, are copper, 401; seawater, 0.62; air at 25°C, 0.026. Water, and seawater in particular, has a value of heat conductivity three or four times greater than wood or wool, but only a fourth that of common rock and, as seen above, far less than that of metals. Seawater or water in general may be said to be a rather poor conductor of heat. In fact the temperature of the ocean falls only by about 20°C from the surface to a depth of 1000 m. Estimating the gradient of temperature to be 0.02°C/m, the downward heat current is approximately $1.2\,\mathrm{W/m^2}$, a quite small value.

Fermi emphasizes the relative smallness of the heat conductivity of seawater by comparing the effects of thermal conductivity in seawater to the actual rate at which heat is transmitted by "mechanical exchanges" such as currents or convection in general. The difference is roughly a factor of 10^5. Using $\kappa_{sw} = 0.62\,\mathrm{W/m\,K}$, and multiplying the heat conductivity of seawater by 10^5, Fermi defines an "effective thermal conductivity," $\kappa_{eff} \sim 6.2 \times 10^4\,\mathrm{W/m\,K}$, or $\kappa_{eff} \sim 143\,\mathrm{cal/cm\,s\,K}$.

[1] The relation to an older commonly used unit is $1\,\mathrm{W/m\,K} = 2.39 \times 10^{-3}\mathrm{cal/cm\,s\,K}$.

Using κ_{eff} instead of κ_{sw} in Eq. (8.1) provides a simple way to estimate the effects of mechanical exchanges. An example of this is given in Sec. (8.3).

8.2 The Heat Equation

Local Heat Absorption In the thermodynamics of a uniform system, the heat transfer and the change in entropy are related by

$$dQ = TdS. \tag{8.2}$$

For a fluid system in local equilibrium, the heat transferred to a system per second is given by the surface integral of the heat current. The rate of heat transfer to a volume of fluid is

$$\frac{dQ}{dt} = -\int_S (\boldsymbol{q} \cdot d\boldsymbol{A}), \tag{8.3}$$

where S is the surface bounding the system, and $d\boldsymbol{A}$ is directed along the outward normal of a surface element. The minus sign in Eq. (8.3) assures that $-(\boldsymbol{q} \cdot d\boldsymbol{A})$ is the rate at which heat flows across $d\boldsymbol{A}$ *into* the volume V. Using the divergence theorem to transform the surface integral to a volume integral, we have

$$\frac{dQ}{dt} = -\int_V (\boldsymbol{\nabla} \cdot \boldsymbol{q}) dV. \tag{8.4}$$

From Eq. (8.4), it follows that $-\boldsymbol{\nabla} \cdot \boldsymbol{q}$ is the local rate of heat transfer/volume. From the second law of thermodynamics, this must equal the product of temperature and the rate of change of the entropy/volume, ρs, where ρ is the mass density, and s is the entropy/mass. We have

$$T\frac{d}{dt}(\rho s) = -\boldsymbol{\nabla} \cdot \boldsymbol{q}. \tag{8.5}$$

In Eq. (8.5) the time derivative for a moving fluid must be the "substantial derivative," or the derivative moving with the fluid, as in Euler's fluid equation. Eq. (8.5) thus becomes

$$T[\frac{\partial(\rho s)}{\partial t} + \boldsymbol{v} \cdot \boldsymbol{\nabla}(\rho s)] = -\boldsymbol{\nabla} \cdot \boldsymbol{q}, \tag{8.6}$$

or

$$T[s(\frac{\partial \rho}{\partial t} + \boldsymbol{v} \cdot \boldsymbol{\nabla}\rho) + \rho(\frac{\partial s}{\partial t} + \boldsymbol{v} \cdot \boldsymbol{\nabla} s)] = -\boldsymbol{\nabla} \cdot \boldsymbol{q}. \tag{8.7}$$

Since ρ obeys the continuity equation, as we see in Eq. (9.2), this simplifies to

$$T\rho(\frac{\partial s}{\partial t} + \boldsymbol{v} \cdot \boldsymbol{\nabla} s) = -\boldsymbol{\nabla} \cdot \boldsymbol{q}. \tag{8.8}$$

Eq. (8.8) ignores viscous effects, but viscosity is negligible for both water and air. It is worth noting that if $\boldsymbol{\nabla} \cdot \boldsymbol{q} = 0$, Eq. (8.8) states that s obeys its own continuity equation; moving with the fluid, the entropy/mass is constant.

Heat Equation The elimination of entropy in Eq. (8.8) takes place after a replacement of the form $Td(\rho s) \rightarrow cdT$, where c is a specific heat/mass for the fluid, in this case seawater. Fermi does not specify whether c means the specific heat at constant volume, c_v, or at constant pressure, c_p. To see why, it is useful at this point to review a few facts about the specific heat of seawater. These are as follows: (1) at all depths of the ocean, the values of c_p and c_v are within 1% of each other; and (2) over the tremendous range of pressures that exist between the surface and bottom of the ocean, c_p and c_v vary by less than 10%. Remarks (1) and (2) also hold accurately in going from pure water to water with the saline content found in ocean water.

So, although the pressure varies over a large range in the ocean, the two common specific heats are essentially equal and insensitive to the pressure. In what follows, we will denote the common specific heat as c_{sw}, assumed constant, with the well-determined surface value $c_{sw} = 4.0\,\mathrm{kJ/kg\,K}$.

Making the replacement $Td(\rho s) \to c_{sw} dT$, and using Eq. (8.1) for $\boldsymbol{q}$, Eq. (8.8) becomes

$$(\rho c)_{sw}(\frac{\partial T}{\partial t} + \boldsymbol{v} \cdot \boldsymbol{\nabla} T) = -\boldsymbol{\nabla} \cdot \boldsymbol{q}. \qquad (8.9)$$

Eq. (8.9) is quite general. In particular, it can be applied when convection and mixing are present. When the heat current is given by Eq. (8.1) and the fluid is at rest, Eq. (8.9) reduces to the familiar form of the heat equation,

$$\frac{\partial T}{\partial t} = \chi_{sw} \nabla^2 T, \qquad (8.10)$$

where

$$\chi_{sw} = (\frac{\kappa}{\rho c})_{sw} \qquad (8.11)$$

is the "thermal diffusivity." Using $\rho_{sw} = 1024\,\mathrm{kg/m^3}, \kappa_{sw} = 0.6\,\mathrm{W/m\,K}$, and $c_{sw} = 4\,\mathrm{kJ/kg\,K}$, we have $\chi_{sw} = 1.47 \times 10^{-7}\,\mathrm{m^2/s}$. The diffusivity is tabulated for many materials. As is obvious from Eq. (8.11), better heat conductors have higher χ values. For example the thermal diffusivity of gold is larger than χ_{sw} by a factor of roughly 1000.

8.3 Diurnal Temperature Variation

We now turn to Fermi's application of Eq. (8.10) to the flow of heat near the surface of the ocean. The heat flow is taken to be perpendicular to the surface of the sea, so we have a single spatial derivative,

$$\frac{\partial T}{\partial t} = \chi_{sw} \frac{\partial^2 T}{\partial z^2}. \tag{8.12}$$

The solution sought has both oscillatory and nonoscillatory time variation. The former has a frequency determined by the length of a day and accounts for the daily variation in heat incident on the ocean's surface. The nonoscillatory terms allow for a mean temperature and mean temperature gradient in the ocean. Taking the positive direction for z downward, a solution of Eq. (8.12) which satisfies these conditions is

$$T(z) = A + Bz + C \exp(-i\omega t) \exp(-az + ibz), \tag{8.13}$$

where A, B, C, a, b are all constant, and we temporarily drop the sw subscript. We will eventually superpose the oscillatory term in Eq. (8.13) and its complex conjugate to obtain a real function. Demanding that the trial form satisfy Eq. (8.12), we have

$$-i\omega = \chi(-a + ib)^2 = \chi(a^2 - 2iab - b^2), \tag{8.14}$$

or

$$\omega = \chi[i(a^2 - b^2) + 2ab]. \tag{8.15}$$

The real and imaginary parts of this equation give

$$a^2 = b^2, \tag{8.16}$$

$$\omega = 2\chi ab.$$

On physical grounds a must be positive. From Eqs. (8.16), we have $|b| = |a|$, with b and ω having the same sign. Using all this, a real form for $T(t, z)$ is

$$T(t, z) = A + Bz + \delta T \cos(\omega t - az + \eta)e^{-az}, \tag{8.17}$$

where η is a constant phase. For the application of interest, ω is determined by the length of a day, 86,400 s, so

$$\omega = \frac{2\pi}{86,400}\ \mathrm{rad/s}, \tag{8.18}$$

and

$$a = \sqrt{\frac{\omega}{2\chi}}. \tag{8.19}$$

Before evaluating a, let us determine A and B. If we average over one day, the oscillatory terms in $T(t,z)$ give zero. Denoting the temperature averaged over a day at sea level by $\bar{T}_0$, and that at depth $z = -d$ by $\bar{T}_d$, A and B are determined. Our formula for $T(t,z)$ becomes

$$T(t,z) = \bar{T}_0 + \frac{\bar{T}_0 - \bar{T}_d}{d} z + \delta T \cos(\omega t - az + \eta) e^{-az}. \tag{8.20}$$

As reasonable values we may take the average ocean temperature at sea level to be 22°C and 4°C at $d = 1000$ m. These would give

$$T(t,z) = 22°\mathrm{C} + (1.8\times 10^{-2}\ \mathrm{C/m}) z + \delta T \cos(\omega t - az + \eta) e^{-az}. \tag{8.21}$$

We will not attempt to find δT quantitatively, but from the variation of air temperature over a day at sea level, a rough estimate of δT is a few C.

Finally, let us determine a. Using Eq. (8.18) and our value of χ_{sw} from above, we have

$$a_{sw} = 15.7\mathrm{m}^{-1}, \ \text{or}\ \frac{1}{a_{sw}} = 6.6\ \mathrm{cm}. \tag{8.22}$$

This result implies that the oscillatory term in the temperature is damped away at quite small depths. For example, if $z = 25$ cm the oscillating term in $T(t,z)$ is reduced to only 2% of its sea level value, and by a depth of 1 m the oscillating term would be too small to detect. In contrast, actual data taken from the ocean show diurnal variations of temperature of approximately 0.15°C at a depth of 5 m. The likely cause of this is that the simple form of heat current given in Eq. (8.1) does not apply. Convection and mixing are certainly present in the layer of the ocean nearest the surface. A remedy for the situation is, following Fermi, to use the effective heat conductivity $\kappa_{eff} \sim 6.2\times10^4$ W/m K. This raises χ_{sw} by a factor of 10^3 and increases $1/a$ by $\sqrt{1000}$, so we have

$$\frac{1}{a_{sw}} \rightarrow \frac{1}{a_{eff}} = (\sqrt{1000})6.6\,\mathrm{cm} = 209\,\mathrm{cm},$$

a step in the right direction. The fact that diurnal temperature variations are observable at a depth of 5 m is a definite indication that the actual heat current is much larger than the simple form of Eq. (8.1) evaluated using κ_{sw}.

For annual variations the a is smaller than the daily value by a factor of $\sqrt{365}$. The diurnal and seasonal variations in temperature are smaller in bodies of water than on dry land, principally because of water's large specific heat.

8.4 Temperature of Seawater

The subject of ocean circulation is a complicated one, ordinarily treated in an oceanography course, but a few remarks along the lines of those articulated by Fermi are appropriate here. The main source of surface water heating is solar radiation, more than twice as great near the equator as near the poles, while the difference in re-radiation is comparatively small. The average surface temperature is approximately 17°C, but temperatures as low as −2°C have been recorded near the poles and as high as 36°C in the Persian

Table 8.1: Average surface temperature versus latitude

Latitude (°N)	0	30	60	90
Temperature(C)	27.1	21.3	4.8	-1.7

Gulf. Temperatures below zero are possible because of seawater's salinity: at the average salinity value of 35 parts per thousand in weight, water does not freeze until -1.94°C. In Table (8.1) we show some average surface temperatures at different latitudes.

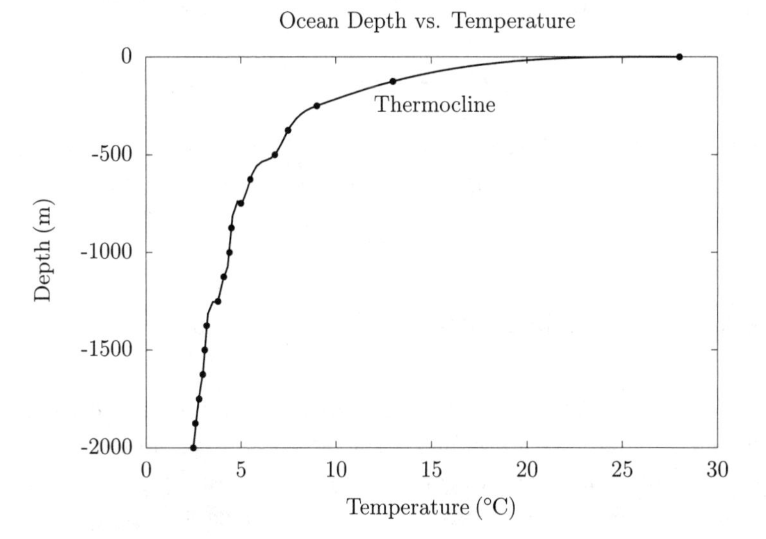

Figure 8.1: Ocean temperature vs. depth
Source: NOAA

Though solar heat is absorbed primarily in the top centimeters of the ocean's surface, winds and waves distribute that surface heat sufficiently to the top 100 m to keep the temperature relatively uniform. Below that level we have a precipitous drop in temperature in a region of rapid temperature change known as a thermocline; it is shown in Fig. (8.1) for a latitude near the equator. Further down, starting at a depth of approximately 1000 m, we reach what is known as the deep ocean, a region containing some 90% of the ocean's volume. From this depth, temperature remains approximately constant, typically at between 0°C and 3°C, with an average estimated at 2.5°C. The thermocline is of course only pronounced in regions of the ocean where the surface temperature is considerably above that of the deep ocean. The amount of heat flowing downward in the ocean may also be estimated by calculating $\kappa_{eff} dT/dz$. Taking a section of the thermocline in which the temperature drops by 1°C in a kilometer, we have

$$\kappa_{eff} dT/dz \approx 0.62\,\mathrm{W/m^2}. \tag{8.23}$$

For deep freshwater, such as found at the bottom of deep lakes, the absence of salts leads to an average temperature of ≈ 4°C.

8.5 Ocean Circulation

As already mentioned, Fermi emphasizes that what he calls "mechanical agitation" is by far the primary factor in the thermal balance of ocean water. The presence of currents with different temperatures has been known for centuries; for example, Benjamin Franklin documented the existence of what we now call the Gulf Stream in a record of ocean temperatures he measured during a crossing of the Atlantic.To understand the existence of currents, salinity also needs to be taken into account.

High-salinity water is denser and tends to sink, causing a sideways movement of less-dense water to replace it. The interplay between temperature and salinity is so important that scientists speak of thermohaline circulation in describing the system of currents that constitute what is sometimes known as the global conveyor belt, a system of sinking and rising masses of water that dictates both the surface and deep ocean flow of water throughout our planet.

(11)

Gravitational waves

$h \gg \frac{\lambda}{2\pi}$ $\quad v = \sqrt{\frac{g\lambda}{2\pi}}$ $\quad \lambda = 200$ m, $v = 18$ m/sec

$h \ll \lambda$ $\quad v = \sqrt{gh}$ $\quad h = 4000$ m, $v = 200$ m/sec

Periods of lakes and oceans

h

a

$\lambda = 2a$

$$\nu = \frac{v}{\lambda} = \frac{\sqrt{gh}}{2a}$$

meters
seconds

$$\nu = \frac{\sqrt{10h}}{2a}$$

$h = 300$ m
$a = 20$ Km

$$T = \frac{2a}{\sqrt{10h}} \qquad T = \frac{20000}{\sqrt{10 \times 300}} = 800 \text{ sec}$$

Period of ocean $\quad a = 4000$ Km

$h = 4000$ m

$$T = \frac{2 \times 4 \times 10^6}{\sqrt{10 \times 4000}} = \frac{8 \times 10^6}{200} = 4 \times 10^4 \sim 12 \text{ hours} \quad 1 \text{ day} = 86400 \text{ sec}$$

Tidal forces

$$A = A_R + A_0 + A_{centrip} = F_{earth} + F_{moon} \qquad \vec{g} = F_{earth} - A_{centrip}$$

$$A_R = g + F_{moon} - A_0$$

Ocean and lake tidal effects

CHAPTER 9

Gravity Waves

Fermi discusses waves that propagate on the surface of oceans and lakes. The fluid in question is water and gravity is the external force acting on the fluid. Such waves are often called "gravity waves." These are not to be confused with the recently detected waves of the same name discovered in 2015. The latter represent oscillations of the very fabric of space and time. The gravity waves which propagate on the surface of water have a counterpart in surface seismic waves, the subject of Chap. (13).

9.1 Dispersion Relations

The relation between frequency f and wavelength λ of any wave is called its "dispersion relation." Rather than a formula for $f(\lambda)$, the relation is often expressed as $\omega(k)$, where $\omega = 2\pi f$, and wave number $k = 2\pi/\lambda$. In either language, the dispersion relation determines the velocity of propagation of the waves. There are actually two velocities. The phase velocity, v_p, is defined by

$$v_p \equiv \frac{\omega}{k} = f\lambda = \frac{\lambda}{T},$$

where T is the period. For an infinitely long wave, the phase velocity specifies the velocity of propagation of surfaces of constant amplitude or "phase." However, infinitely long waves with definite frequency and wavelength are an idealization. A physically realizable wave is constructed by superposing a continuum of wavelengths near a dominant wavelength. Such wave packets, as they are called, move with the group velocity, v_g, defined by

$$v_g \equiv \frac{\partial \omega}{\partial k} = \frac{\partial f}{\partial(1/\lambda)}. \tag{9.1}$$

9.2 Euler's Equation and the Equation of Continuity

The equations of fluid dynamics are macroscopic equations, replacing a microscopic description at the molecular level with a small number of average quantities: (1) the mass density ρ; (2) the local velocity $\boldsymbol{v}$; and (3) the pressure p. This is similar to what is done in thermodynamics, but variation of quantities in space and time are allowed in fluid dynamics.

For the analysis of surface waves in water, one of the two fluid equations needed is the equation of continuity,

$$\frac{\partial \rho}{\partial t} + \boldsymbol{\nabla} \cdot (\rho \boldsymbol{v}) = 0, \tag{9.2}$$

which states that water is neither created nor destroyed. It is very similar to the equation in electrodynamics stating that electric charge is neither created nor destroyed.

The other equation needed is Euler's equation,

$$\rho(\frac{\partial \boldsymbol{v}}{\partial t} + \boldsymbol{v} \cdot \boldsymbol{\nabla} \boldsymbol{v}) = -\boldsymbol{\nabla} p - \rho \boldsymbol{\nabla} V_g, \tag{9.3}$$

where V_g is the gravitational potential. For waves traveling on the surface of the ocean or a lake, V_g is given by the elementary formula $V_g = gz$, where z is the local vertical coordinate.[1] Euler's equation is the expression of Newton's second law for a fluid. The right-hand side of Eq. (9.3) is the force/volume acting on the fluid. The term $-\boldsymbol{\nabla} p$ is the force/volume produced by the spatial variation of pressure. A difference of pressure on opposite sides of a small volume of fluid produces a net force in the direction of decreasing pressure. The term in V_g is the force/volume exerted on water by the Earth's gravity. Water is accurately described as an "ideal fluid," meaning viscous forces are negligible, so there are no other terms on the right side of Eq. (9.3). The left side of Euler's equation is the rate of change of momentum/volume of the fluid. In Newton's second law, as applied to the motion of a projectile, the acceleration is evaluated along the trajectory of the projectile. The same must hold for a small volume of fluid. Such a volume may have a velocity at time t of $\boldsymbol{v}(\boldsymbol{x}, t)$. At an instant dt later, the velocity of this small volume including its motion is

$$\boldsymbol{v}(\boldsymbol{x}+\boldsymbol{v}dt, t+dt) = \boldsymbol{v}(\boldsymbol{x}, t) + (\frac{\partial \boldsymbol{v}}{\partial t}(\boldsymbol{x}, t) + \boldsymbol{v} \cdot \boldsymbol{\nabla} \boldsymbol{v}(\boldsymbol{x}, t))dt. \tag{9.4}$$

From this, we see that the rate of change of $\boldsymbol{v}$ moving with the fluid is

$$\left(\frac{\partial \boldsymbol{v}}{\partial t}(\boldsymbol{x}, t) + \boldsymbol{v} \cdot \boldsymbol{\nabla} \boldsymbol{v}(\boldsymbol{x}, t)\right). \tag{9.5}$$

The presence of the $\boldsymbol{v} \cdot \boldsymbol{\nabla} \boldsymbol{v}$ term makes Euler's equation nonlinear and very difficult to solve. We will proceed by ignoring

[1] Note that in this section the z axis points upward.

quadratic terms in the velocity, and define the conditions for which this is valid after the solution of the linearized equation has been found.

9.3 Small Amplitude Surface Waves

Water occupies the region $z < 0$, with the atmosphere in the region $z > 0$. We are seeking a surface wave, one whose amplitude is largest at the water surface, damped when moving deeper into the water. As is typical for most fluid waves, there are no vortices present, so $\boldsymbol{\nabla} \times \boldsymbol{v} = 0$. This allows the velocity to be derived from a velocity potential,

$$\boldsymbol{v} = \boldsymbol{\nabla}\phi. \tag{9.6}$$

Treating the density of water as constant, and using the continuity equation, Eq. (9.2), we also have $\boldsymbol{\nabla} \cdot \boldsymbol{v} = 0$. The vanishing of $\boldsymbol{\nabla} \cdot \boldsymbol{v}$ along with Eq. (9.6) implies the Laplace equation for ϕ,

$$\boldsymbol{\nabla} \cdot \boldsymbol{v} = 0 \rightarrow \nabla^2\phi = 0. \tag{9.7}$$

Euler's equation (for the case where quadratic terms in the velocity can be neglected) is

$$\rho\frac{\partial \boldsymbol{v}}{\partial t} = -\boldsymbol{\nabla}(p + \rho g z). \tag{9.8}$$

As discussed explicitly below, ignoring quadratic terms in the velocity is justified if the wavelength of the wave is large compared to the amplitude of the wave.

Using Eq. (9.6) and the constancy of ρ to rewrite Eq. (9.8), we have

$$\rho\frac{\partial}{\partial t}\boldsymbol{\nabla}\phi = \boldsymbol{\nabla}(\rho\frac{\partial \phi}{\partial t}) = -\boldsymbol{\nabla}(p + \rho g z). \tag{9.9}$$

It is desirable to remove the gradient operator in Eq. (9.9) and get an equation relating the time derivative of the velocity potential directly to the pressure and gravitational potential. To proceed, consider first the case where the fluid is at rest and $\phi \equiv 0$. Then we have

$$\nabla(p + \rho g z) = 0. \tag{9.10}$$

If the surface of the water is at $z = 0$, the pressure is the atmospheric pressure, p_a. The solution to Eq. (9.10) is given by the barometric formula

$$p_b(z) + \rho g z = p_a. \tag{9.11}$$

Returning to Eq. (9.9), we will take ϕ to be a traveling wave which satisfies Laplace's equation in spatial variables. Removing the gradient, we now write

$$-\rho \frac{\partial \phi}{\partial t} = p - p_b = p - p_a + \rho g z. \tag{9.12}$$

Eq. (9.12) builds in the barometric formula when the amplitude of the traveling wave is reduced to zero. We take our surface wave to be moving in the x direction with wave number k and frequency ω. As a trial form, we set

$$\phi = A \cos(kx - \omega t) e^{kz}. \tag{9.13}$$

Computing second derivatives, we have

$$\frac{\partial^2 \phi}{\partial x^2} = -k^2 \phi, \tag{9.14}$$

$$\frac{\partial^2 \phi}{\partial z^2} = k^2 \phi,$$

so Laplace's equation is satisfied. Writing the exponential factor in ϕ as $\exp(2\pi z/\lambda)$, we see that the penetration depth is $\lambda/2\pi$, so longer wavelengths penetrate more deeply into the ocean.[2]

Returning to Eq. (9.12), we have

$$-\rho\omega A\sin(kx-\omega t)e^{kz}=p-p_a+\rho gz. \tag{9.15}$$

When the wave is present, the surface of the water is no longer at $z=0$, but at a variable distance, $z=\xi(x,t)$. The pressure at the surface is of course still atmospheric. At $z=\xi(x,t)$ and atmospheric pressure, Eq. (9.15) becomes

$$-\rho\omega A\sin(kx-\omega t)e^{k\xi}=\rho g\xi. \tag{9.16}$$

This equation determines the wave amplitude $\xi(x,t)$. However, there is no need to solve it exactly. We are interested in waves of small amplitude, i.e., amplitude much less than the wavelength. Since $k\xi=2\pi\xi/\lambda$, we have $k\xi<<1$, so the factor $\exp(k\xi)\sim 1$. Setting $\exp(k\xi)=1$, Eq. (9.16) now gives an explicit formula for $\xi(x,t)$. Taking the time derivative of ξ gives the z component of velocity at the water's surface,

$$\frac{\partial\xi}{\partial t}=(v_z)_{surface}=\frac{\omega^2}{g}A\cos(kx-\omega t). \tag{9.17}$$

But v_z at any point in the water or on its surface is given by

$$v_z=\frac{\partial\phi}{\partial z}.$$

Taking the derivative with respect to z of Eq. (9.13), we have

$$v_z=kA\cos(kx-\omega t)e^{kz}. \tag{9.18}$$

Evaluating this at $z=\xi$, again setting $\exp(k\xi)=1$, gives a second formula for the z component of velocity at the surface,

$$(v_z)_{surface}=kA\cos(kx-\omega t). \tag{9.19}$$

[2] In Appendix C, we analyze a surface wave that travels in the Earth's crust, where again the penetration depth is controlled by the wavelength.

Equating the two expressions gives

$$\frac{\omega^2}{g} A \cos(kx - \omega t) = kA \cos(kx - \omega t), \tag{9.20}$$

which gives the dispersion relation

$$\omega^2 = kg. \tag{9.21}$$

The phase velocity is

$$v_p = \frac{\omega}{k} = \sqrt{\frac{g}{k}} = \sqrt{\frac{g\lambda}{2\pi}}. \tag{9.22}$$

The group velocity is 1/2 the phase velocity,

$$v_g = \frac{\partial \omega}{\partial k} = \frac{1}{2}\sqrt{\frac{g}{k}} = \frac{1}{2}\sqrt{\frac{g\lambda}{2\pi}}. \tag{9.23}$$

In the present approximation, discussed in the next section, surface waves can be superposed, interfere, scatter off islands, etc. Much more detail on these topics is given in Kinsman's book (Kinsman 1965).

Criterion for Dropping Quadratic Terms in Fluid Velocity When is it valid to drop quadratic terms in the fluid velocity? If quadratic terms are to be ignored, we must have

$$\frac{\partial \boldsymbol{v}}{\partial t} >> \boldsymbol{v} \cdot \boldsymbol{\nabla} \boldsymbol{v}. \tag{9.24}$$

For a wave with frequency ω and wave vector $\boldsymbol{k}$, this condition becomes

$$\omega \boldsymbol{v} >> \boldsymbol{v} \cdot \boldsymbol{k} \boldsymbol{v}. \tag{9.25}$$

Canceling a factor of $\boldsymbol{v}$, and dividing by k, Eq. (9.25) becomes

$$v_p = \frac{\omega}{k} >> |\boldsymbol{v}|. \tag{9.26}$$

This says that the phase (or group) velocity should be large compared to the velocity of the particles in the fluid. For particles at the surface of the fluid, the velocity is $\sim kA \sim A/\lambda$, so dropping quadratic terms is justified if

$$\frac{A}{\lambda} << \sqrt{g\lambda}. \tag{9.27}$$

For a fixed A, this criterion will be satisfied for sufficiently large wavelength λ, or for a fixed wavelength, it will be satisfied for sufficiently small A. As a numerical example, suppose $\lambda = 10\,\mathrm{m}$. For this wavelength, $v_p \sim 4\,\mathrm{m/s}$. Then $A/\lambda << v_p$ will be satisfied for A/λ in the range of $0.4\,\mathrm{m/s}$ or smaller.

Surface Waves in Water of Finite Depth The preceding discussion described the case where the depth of the body of water being considered was effectively infinite. The case of water of finite depth, h, can be treated by a simple generalization of the trial form of Eq. (9.13). For a finite depth of water, both $\exp(kz)$ and $\exp(-kz)$ are allowed in the velocity potential. The boundary condition at the bottom of the body of water is that the velocity normal to the surface must vanish, i.e., $v_z = 0$. A trial form that satisfies the Laplace equation is

$$\phi = A\frac{\cosh[k(z+h)]}{\cosh(kh)}\cos(kx - \omega t). \tag{9.28}$$

The z component of velocity is

$$v_z = \frac{\partial \phi}{\partial z} = Ak\frac{\sinh[k(z+h)]}{\cosh(kh)}\cos(kx - \omega t). \tag{9.29}$$

This vanishes at $z = -h$, so the boundary condition at the bottom of the body of water is satisfied. The rest of the calculation proceeds in a similar manner to the case of unlimited depth. Eq. (9.12) applies here as well. Taking the time derivative of ϕ, the generalization of Eq. (9.15) is

$$-\rho\omega A \frac{\cosh[k(z+h)]}{\cosh(kh)} \sin(kx-\omega t) = p - p_a + \rho g z. \quad (9.30)$$

We have atmospheric pressure at the water's surface, and $z = \xi(x,t)$. At the surface, Eq. (9.30) reduces to

$$-\rho\omega A \frac{\cosh[k(\xi+h)]}{\cosh(kh)} \sin(kx-\omega t) = \rho g \xi. \quad (9.31)$$

Ignoring ξ compared to h, we obtain an explicit formula for ξ,

$$\xi(x,t) = -\frac{\omega A}{g} \sin(kx-\omega t). \quad (9.32)$$

The time derivative of ξ gives $(v_z)_{surface}$. Equating this to $(v_z)_{surface}$ obtained from Eq. (9.29) gives

$$\frac{\omega^2 A}{g} \cos(kx-\omega t) = Ak \tanh(kh) \cos(kx-\omega t), \quad (9.33)$$

where we again ignore ξ compared to h. Dividing common factors on both sides, the dispersion relation for surface gravity waves in water of finite depth is

$$\omega^2 = gk \tanh(kh). \quad (9.34)$$

We see that Eq. (9.34) reduces to Eq. (9.21) when $kh >> 1$, or $h >> \lambda/2\pi$. Actually, since $\tanh(2\pi h/\lambda)$ approaches unity rapidly, Eq. (9.34) is essentially exact once $h > \lambda/2$.

Qualitative Behavior One of the most interesting features of surface gravity waves is the nonlinear relationship between wavelength and period. Taking the case where $h > \lambda/2$ and Eq. (9.21) is valid, we have

$$\lambda = \frac{g}{2\pi} T^2, \quad (9.35)$$

Table 9.1: Wavelengths and phase velocities for surface gravity waves

T(s)	λ(m)	v_p(m/s)
1	1.56	1.56
10	156	15.6
100	15,600	156
1000	1.56×10^6	1560

where $\lambda = 2\pi/k$ is the wavelength and $T = 2\pi/\omega$ the period. In Table (9.1), we show the period, wavelength, and phase velocity for various periods. The behavior seen in the table is quite different from waves such as sound and light waves, where the wavelength grows linearly with T, and the phase velocity is independent of T.

Table (9.1) is for the case $h > \lambda/2$, where Eq. (9.21) holds. For waves on the open ocean, taking a typical depth of $h = 4000\,\text{m}$, such waves will not "feel the bottom" for $\lambda < 8000\,\text{m}$ or $T < 72\,\text{s}$. Now suppose a wave of period $T < 72\,\text{s}$ moves from the open ocean, with wavelength given by Eq. (9.35), into shallow waters, where $\lambda < h/2$ is no longer satisfied. Using Eq. (9.34), we have

$$\lambda = \frac{g}{2\pi} T^2 \tanh(\frac{2\pi h}{\lambda}). \tag{9.36}$$

From this equation, we can see that if h is significantly less than λ, the wavelength will be smaller than its large h value given by Eq. (9.35), so for a given period of the wave, the wavelength decreases when the wave moves from the open ocean toward the shore. As an example, suppose $T = 10\,\text{s}$. In the deep ocean, this corresponds to a wavelength $\lambda = 156\,\text{m}$. Moving to shallower water, the wavelength is reduced to $\sim 70\,\text{m}$ at a depth of $h = 12\,\text{m}$.

Fermi considers two situations: that of a wave sweeping

across either a lake or an ocean, both involving $h \ll \lambda$ and therefore both making use of Eq. (9.36). For $h << \lambda$, Eq. (9.36) simplifies to

$$\lambda = \sqrt{gh}T, \tag{9.37}$$

so in this limit, surface waves have velocity independent of period. For a lake of depth $h = 300\,\mathrm{m}$ and width $a = 20\,\mathrm{km}$, a wave with a maximum on one side and a minimum on the other side will have $\lambda = 2a = 4 \times 10^4\,\mathrm{m}$. Eq. (9.37) gives $T = 738\,\mathrm{s}$.

For the analogous case of an ocean tide, Fermi chooses $h \sim 4000\,\mathrm{m}$, a depth lying roughly in between the average depths of the Pacific and Atlantic Oceans, and as width he chooses $a \sim 4000\,\mathrm{km}$. In this case the period for the disturbance is

$$T \sim 4 \times 10^4\,\mathrm{s},$$

a little less than 12 hours. The ocean, or any body of water, has its own characteristic modes of oscillation. In addition, ocean tides are driven by the gravitational forces of the Sun and Moon. This is discussed in detail in the next chapter.

$$A_{0x} = -\frac{GM}{r_0^2} \qquad F_{Mx} = -\frac{GM}{r_0^2} + \frac{2GM}{r_0^3}x$$

$$A_{0y} = 0 \qquad F_{My} = -\frac{GM}{r_0^3}y$$

$$A_{0z} = 0 \qquad F_{Mz} = -\frac{GM}{r_0^3}z$$

Moon

Earth

Tidal forces $\frac{GM}{r_0^3}\begin{bmatrix} 2x \\ -y \\ -z \end{bmatrix}$

$M_{moon} = 7.34 \times 10^{25}$ $\quad r_{0\,moon} = 3.84 \times 10^{10}$ $\quad G = 6.67 \times 10^{-8}$

$M_{sun} = 2 \times 10^{33}$ $\quad r_{0\,sun} = 1.49 \times 10^{13}$

$$k = \frac{GMa}{r_0^3} = \frac{6.67 \times 10^{-8} \times 7.34 \times 10^{25} \times 6.37 \times 10^{8}}{3.84^3 \; 10^{30}} = 5.5 \times 10^{-5} \quad \text{Moon}$$

$$k = \frac{GMa}{r_0^3} = \qquad = 2.6 \times 10^{-5} \quad \text{Sun}$$

Tidal periods 12.42 h moon
12.00 h sun

Maximum tides (new or full moon)

Tidal potential

$$-\frac{GMa}{r_0^3}\left(x^2 - \frac{y^2+z^2}{2}\right) = -\frac{GMa^2}{2r_0^3}\left(3\cos^2\alpha - 1\right)$$

Equilibrium shape

$$-\frac{ga^2}{r} - \frac{GMa^2}{2r_0^3}\left(3\cos^2\alpha - 1\right) = \text{const}$$

$$\delta r = \frac{GMa^2}{3r_0^3 g}\left(3\cos^2\alpha - 1\right) = 36 \text{ cm} \left(3\cos^2\alpha - 1\right)$$

Tidal forces caused by the Moon

CHAPTER 10

Tide Physics

10.1 Effects of the Moon

The Moon exerts an overall force of attraction on the Earth. According to Newton's law of gravity, this force is given by

$$\boldsymbol{F}_{em} = \frac{GM_m M_e}{R^3}\boldsymbol{R}, \tag{10.1}$$

where $\boldsymbol{R}$ is the vector from the Earth's center of mass to the Moon.[1] In addition to this total force, the Moon exerts what are known as "tidal forces" on the Earth. To define exactly what a tidal force is, consider the gravitational force of the Moon on a small mass δ, which is located at a position $\boldsymbol{r}$, either in the interior of the Earth or on its surface.

The vector pointing to the Moon from the location of δ is

[1] The Moon will be treated as a point mass in what follows.

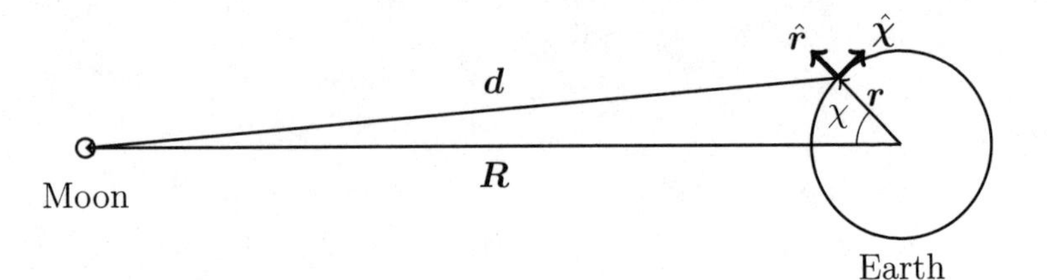

Figure 10.1: The Earth-Moon system

$\boldsymbol{d} = \boldsymbol{R} - \boldsymbol{r}$. Again using Newton's law, the force on δ is

$$\boldsymbol{F}_{\delta m}(\boldsymbol{r}) = \frac{GM_m\delta}{d^3}\boldsymbol{d} = \frac{GM_m\delta}{|\boldsymbol{R}-\boldsymbol{r}|^3}(\boldsymbol{R}-\boldsymbol{r}). \tag{10.2}$$

Extracting the factor of δ in Eq. (10.2), we can define the Moon's gravitational acceleration at $\boldsymbol{r}$ by

$$\boldsymbol{g}_m(\boldsymbol{r}) \equiv \frac{GM_m}{|\boldsymbol{R}-\boldsymbol{r}|^3}(\boldsymbol{R}-\boldsymbol{r}). \tag{10.3}$$

At $\boldsymbol{r} = 0$, $\boldsymbol{g}_m(\boldsymbol{r})$ reduces to

$$\boldsymbol{g}_m(\boldsymbol{0}) = \frac{GM_m}{|\boldsymbol{R}|^3}(\boldsymbol{R}). \tag{10.4}$$

Comparing to Eq. (10.1), we see that

$$\boldsymbol{F}_{em} = \boldsymbol{g}_m(\boldsymbol{0})M_e. \tag{10.5}$$

It is clear from Eq. (10.5) that $\boldsymbol{g}_m(\boldsymbol{0})$ is the acceleration due to the Moon that the entire Earth experiences (Stacey 1969). Every part of the Earth's mass, including the small mass δ, experiences this acceleration. The tidal acceleration is defined to be the difference between $\boldsymbol{g}_m(\boldsymbol{r})$ and $\boldsymbol{g}_m(\boldsymbol{0})$,

$$\boldsymbol{g}_{tm}(\boldsymbol{r}) \equiv \boldsymbol{g}_m(\boldsymbol{r}) - \boldsymbol{g}_m(\boldsymbol{0}). \tag{10.6}$$

Using Eqs. (10.2), (10.3), and (10.4), we can express the Moon's force on the small mass δ as

$$\boldsymbol{F}_{\delta m}(\boldsymbol{r}) = \boldsymbol{g}_{tm}(\boldsymbol{r})\delta + \boldsymbol{g}_m(\boldsymbol{0})\delta. \tag{10.7}$$

The second term in Eq. (10.7) is just the contribution of δ to the total force exerted by the Moon on the Earth. The first term, $\boldsymbol{g}_{tm}(\boldsymbol{r})\delta$, is the tidal force on δ. By acting differently on its constituent parts, tidal forces in general tend to change the shape of a massive body.

Substituting from Eqs. (10.3) and (10.4), we have

$$\boldsymbol{g}_{tm}(\boldsymbol{r}) = \frac{GM_m}{|\boldsymbol{R}-\boldsymbol{r}|^3}(\boldsymbol{R}-\boldsymbol{r}) - \frac{GM_m}{R^3}\boldsymbol{R}. \tag{10.8}$$

Since the ratio of the Earth's radius to the Earth-Moon distance is small (1.66×10^{-2}), it is a good approximation to evaluate $g_{tm}(\boldsymbol{r})$ to first order in r/R. Expanding the factor $1/|\boldsymbol{R}-\boldsymbol{r}|^3$, we have

$$\frac{1}{|\boldsymbol{R}-\boldsymbol{r}|^3} = \frac{1}{R^3}[1 - \frac{2r\cos\chi}{R} + (\frac{r}{R})^2]^{-3/2} \tag{10.9}$$
$$= \frac{1}{R^3}(1 + \frac{3r\cos\chi}{R} + \ldots).$$

Substituting this expression into Eq. (10.8) and keeping only terms of order r/R, we have

$$\boldsymbol{g}_{tm}(\boldsymbol{r}) = \frac{GM_m}{R^3}\left(\frac{3r\cos\chi}{R}\boldsymbol{R} - \boldsymbol{r}\right). \tag{10.10}$$

It is useful to break $\boldsymbol{R}$ into components tangent and perpendicular to the Earth at $\boldsymbol{r}$. In terms of the unit vectors $\hat{\boldsymbol{r}}$ and $\hat{\boldsymbol{\chi}}$ (see Fig. (10.1)), $\boldsymbol{R}$ is

$$\boldsymbol{R} = R(\hat{\boldsymbol{r}}\cos\chi - \hat{\boldsymbol{\chi}}\sin\chi). \tag{10.11}$$

Substituting this expression into Eq. (10.10), $\boldsymbol{g}_{tm}(\boldsymbol{r})$ becomes

$$\boldsymbol{g}_{tm}(\boldsymbol{r}) = \frac{GM_m}{R^2}(\frac{r}{R})\left(\hat{\boldsymbol{r}}(3\cos^2\chi - 1) - \hat{\boldsymbol{\chi}}\, 3\sin\chi\cos\chi\right). \tag{10.12}$$

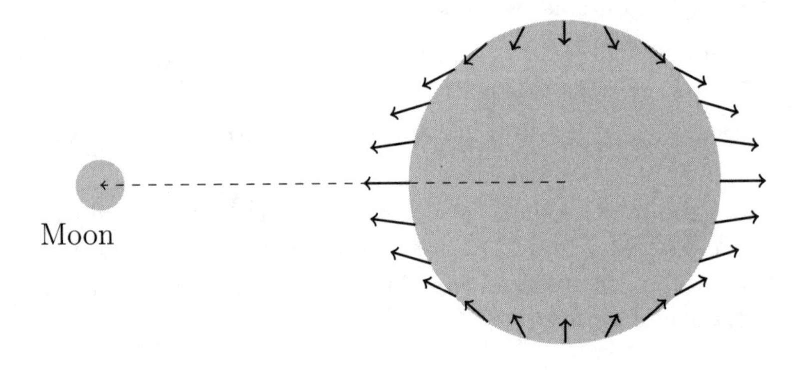

Figure 10.2: Tidal forces on the Earth's surface

Fig. (10.2) is an arbitrary section through the center of the Earth and shows the pattern of tidal forces derived from $\boldsymbol{g}_{tm}(\boldsymbol{r})$ acting at the Earth's surface. On the hemisphere closest to the Moon, tidal forces tend to push matter toward the point which lies nearest to the Moon. On the hemisphere further from the Moon, the pattern repeats with tidal forces pushing matter to the point farthest from the Moon. The tidal force tangent to the Earth has maximal magnitude at $\chi = \pi/4$ and $\chi = 3\pi/4$.

10.2 The Tidal Potential

The tidal acceleration due to the Moon can be derived from a tidal potential, as W_{tm},

$$\boldsymbol{g}_{tm}(\boldsymbol{r}) = -\boldsymbol{\nabla} W_{tm}(\boldsymbol{r}) = -(\hat{\boldsymbol{r}}\frac{\partial}{\partial r} + \hat{\boldsymbol{\chi}}\frac{1}{r}\frac{\partial}{\partial \chi})W_{tm}(\boldsymbol{r}). \quad (10.13)$$

It is easy to check that taking the negative gradient of

$$W_{tm}(\boldsymbol{r}) = -\frac{GM_m r^2}{2R^3}(3\cos^2\chi - 1) \quad (10.14)$$

reproduces Eq. (10.12) for $\boldsymbol{g}_{tm}(\boldsymbol{r})$. Regarded as a function of χ, $W_{tm}(\boldsymbol{r})$ has minima at $\chi = 0$ and $\chi = \pi$. These correspond to the points on the Earth nearest to and farthest from the Moon in Fig. (10.2).

10.3 Surface Displacement for an Earth Covered with Water

A large fraction of the Earth's surface (71%) is covered with water. To gain insight into how the Moon affects the distribution of water on the Earth, we consider an idealized situation in which the Earth is taken to be a solid nonrotating sphere, covered entirely by ocean whose depth is uniform in the absence of the Moon.

Once the Moon's tidal forces are taken into account, the depth of the ocean will vary from point to point, depending on the location of the point being considered relative to the Moon. The magnitude of this variation follows from the requirement that the surface of the ocean be a gravitational equipotential, so that there is no net force tangent to the liquid surface. The calculation to follow is quite similar to the one which determined $r(\theta)$, in the discussion of Clairaut's theorem in Sec. (2.2) (see Eq. (2.25).

In the absence of the Moon, the Earth's radius a is the distance from the surface of the ocean to the Earth's center of mass. In the presence of the Moon and its accompanying tidal forces, the liquid surface of the Earth will be distorted by a small amount we denote as η. At a point on the surface, the distance from the center of the Earth is now $a+\eta$ instead of a. The total gravitational potential at such a point receives a contribution from the Earth as well as the Moon. The total gravitational potential at the surface is

$$V_{tot} = -\frac{GM_e}{(a+\eta)} + W_t(\chi). \tag{10.15}$$

Working to first order in the small quantities η and $|W_t|$, we may set $r = a$ in evaluating W_t. The tidal potential then depends only on the angle χ defined in Fig. (10.1). In order for V_{tot} to be an equipotential at the surface, it cannot depend on the angle χ. This can be accomplished by allowing η to depend on χ. Expanding V_{tot} to first order, we have

$$V_{tot} = -\frac{GM_e}{a}(1 - \frac{\eta}{a}) - \frac{GM_m}{2R}(\frac{a}{R})^2(3\cos^2\chi - 1). \tag{10.16}$$

Requiring that the term linear in η cancel the contribution of the tidal potential, we have

$$\frac{\eta}{a} = \frac{M_m}{M_e}(\frac{a}{R})^3\frac{1}{2}(3\cos^2\chi - 1) + C, \tag{10.17}$$

where the constant C is independent of χ. The value of C is determined by the requirement that the distortion of the surface away from purely spherical cannot change $\mathcal{V}_{oc}$, the total volume of water in the ocean. The change in the volume of the ocean can be written to first order as

$$\Delta\mathcal{V}_{oc} = \int \eta a^2 d\Omega, \tag{10.18}$$

where $d\Omega$ is the element of solid angle. The term in $\Delta\mathcal{V}_{oc}$ arising from the tidal potential integrates to zero, leaving

$$\Delta\mathcal{V}_{oc} = 4\pi a^2 C. \tag{10.19}$$

Requiring that $\Delta\mathcal{V}_{oc} = 0$ then implies $C = 0$. We are left with

$$\eta = a(\frac{M_m}{M_e})(\frac{a}{R})^3 \frac{1}{2}(3\cos^2\chi - 1). \tag{10.20}$$

Using the known values for the parameters in Eq. (10.20), we have for the points on the Earth which are nearest and farthest from the Moon

$$\eta(0) = \eta(\pi) \approx 36\,\text{cm}, \tag{10.21}$$

while for $\chi = \pi/2$,

$$\eta(\pi/2) \approx -18\,\text{cm}. \tag{10.22}$$

A vessel traveling on this all-ocean Earth engaged in depth sounding would find the ocean to be 36 cm deeper at $\chi = 0$ or π, and 18 cm shallower at $\chi = \pi/2$. The terms "deeper" and "shallower" are a comparison of the distance from the surface of the ocean to the bottom of the ocean in the presence of the Moon's tidal forces vs. the case where the Moon's tidal forces are absent.

The situation so far is purely static, whereas tides are intrinsically time-dependent phenomena. Actual tides can be introduced on our model ocean-covered Earth by allowing the Earth to rotate. From an Earth-fixed viewpoint, the Moon will appear to rotate around the Earth. Referring to Fig. (10.2), for simplicity ignore the Earth's tilt angle so the Earth's spin axis is perpendicular to the page. As the Moon rotates around the Earth, the gravitational potential from the Moon will be time-dependent, since now the angle $\chi = \omega t$, where $\omega = 2\pi/T_{day}$. A small island on the equator would experience the angles $\chi = 0, \pi/2, \pi, \pi/2, 0$ in succession with "high tides" at $\chi = 0, \pi$, and "low tides" at $\chi = \pi/2, \pi/2$.

Returning to the actual Earth, the maximum and minimum values of η given above can be used to make order of magnitude estimates of the height of actual tides in many places. Much larger values do occur. In particular the shape of the ocean basin can act to greatly amplify the magnitude of tides.

10.4 Time Dependence of Tides

It is familiar that each day there are two high tides and two low tides. Both are easily experienced at any point along an ocean's shore. These tides are caused by tidal forces from the Moon and also the Sun. What is meant by a "day"is slightly different in the two cases. For solar tides, a day is simply 24 hours, denoted from here on by T_s. For lunar tides, the day is the "lunar day" which is 24 hours and 50 minutes, denoted in the following as T_l. The lunar day is longer than the solar day because the moon has moved a small but perceptible amount in 24 hours. This is discussed in more detail in ndix A. After the passage of a lunar day, a point on the Earth will return to a position whose orientation to the Moon is very close to the original orientation.[2]

Fig. (10.3) shows tidal data observed at a point on the eastern coast of Florida. Although the time scale shown in the figure does not allow the difference between T_l and T_s to be resolved, it does clearly show that there are two high tides and two low tides in a period of roughly 24 hours. Further,

[2]The tilt of the Moon's orbit prevents the two orientations from being exactly the same.

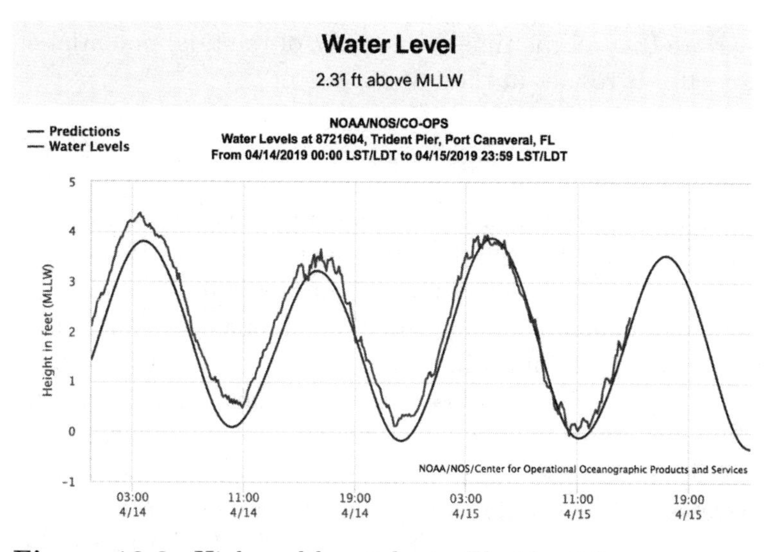

Figure 10.3: High and low tides in Florida, US
Source: NOAA

the high tides spaced by approximately 12 hours clearly have different amplitudes.

Lunar Tide Time Dependence The goal of this section is to understand why lunar high tides spaced by $T_l/2$ have different amplitudes.[3] Start by imagining that the plane of the Moon's orbit coincides with the plane of the Earth's equator. Then in both Fig. (10.1) and Fig. (10.2), the spin axis of the Earth is perpendicular to the page. Fig. (10.2) makes it clear that as the Earth rotates around its spin axis, the Earth will return to the same orientation with respect to the Moon after a half lunar day. For this case, the high tides spaced apart by half a lunar day would have the same amplitude.

[3]Additional tidal effects from the Moon that depend on the length of the lunar month will not be discussed here.

Note that all the time dependence of the tidal potential of Eq. (10.14) resides in the way the factor

$$\cos^2\chi = \frac{1}{2}(1+\cos(2\chi)) \qquad (10.23)$$

changes as the Earth rotates. To see how $\cos^2\chi$ varies when the Moon is not in the Earth's equatorial plane, it is useful to adopt a coordinate system with the origin at the Earth's center of mass, the z axis along the Earth's spin axis, and the x and y axes fixed in orientation with respect to distant stars (Cook 1973). In this system, the Earth rotates around the z axis, the x–y plane being the plane of the equator. The vector $\boldsymbol{R}$ from the origin to the Moon can be represented by its spherical coordinates, R, θ_m, ϕ_m in this coordinate system. Likewise the vector $\boldsymbol{r}$ to a point on or in the Earth can be represented by r, θ_e, ϕ_e. Using standard formulas, the unit vectors $\hat{\boldsymbol{r}}$ and $\hat{\boldsymbol{R}}$ are

$$\hat{\boldsymbol{r}} = \cos\theta_e\hat{\boldsymbol{z}} + \sin\theta_e(\hat{\boldsymbol{x}}\cos\phi_e + \hat{\boldsymbol{y}}\sin\phi_e), \qquad (10.24)$$

and

$$\hat{\boldsymbol{R}} = \cos\theta_m\hat{\boldsymbol{z}} + \sin\theta_m(\hat{\boldsymbol{x}}\cos\phi_m + \hat{\boldsymbol{y}}\sin\phi_m).$$

Taking the scalar product of these two unit vectors gives $\cos\chi$. We have

$$\cos\chi = \cos\theta_e\cos\theta_m + \sin\theta_e\sin\theta_m\cos(\phi_e - \phi_m). \qquad (10.25)$$

Having accounted for the most important part of the Moon's movement by using the lunar day, the only time dependence in Eq. (10.25) comes from the Earth's rotation, which causes ϕ_e to vary in time as

$$\phi_e(t) = \phi_e(0) + \frac{2\pi t}{T_l}. \qquad (10.26)$$

As mentioned above, the tidal potential depends on the square of $\cos\chi$. Without writing out the details, it is clear from

Eq. (10.25) that when we compute $\cos^2\chi$, there will be a term in $\cos(\phi_e - \phi_m)$, and one in $\cos^2(\phi_e - \phi_m)$. The first of these is periodic with a period of T_l, while the second is periodic with period $T_l/2$. This shows that the tidal potential from the Moon, and therefore the Moon's tidal forces, contains terms of two types: one periodic with a period of $T_l/2$, the other periodic with a period T_l. We will not attempt to actually calculate the differences between the high tides at these two intervals, but the discussion just given makes it clear that in general they will be different. Note that putting the Moon back in the Earth's equatorial plane amounts to setting $\cos\theta_m = 0$, and then $\cos^2\chi$ becomes periodic with period $T_l/2$.

Solar Tide Time Dependence The analysis just given for the effects of the Moon on Earth tides can repeated for the effect of the Sun on Earth tides. Rewriting Eq. (10.20) for the Sun, we have

$$\eta(\chi_s) = r(\frac{M_s}{M_e})(\frac{r}{R_{e-s}})^3\frac{1}{2}(3\cos^2\chi_s - 1), \tag{10.27}$$

where χ_s is measured relative to the vector from the Earth's center of mass to the Sun. Evaluating η for $\chi_s = 0$ or π, we have

$$\eta_s(\chi_s = 0) = \eta_s(\chi_s = \pi) = 16.3\,\text{cm}, \tag{10.28}$$

approximately half the effect from the Moon. In general, the Sun and Moon are not aligned, so their net effect is a vector sum,

$$\boldsymbol{\eta}(\chi_m) + \boldsymbol{\eta}(\chi_s). \tag{10.29}$$

There is a contribution to the tidal potential due to the Sun of similar form to Eq. (10.14). In general the ratio of tidal effects from the Sun to those from the Moon goes as

$$(\frac{M_s}{M_m})(\frac{R_{e-m}}{R_{e-s}})^3 \approx 0.45,$$

so although the tidal effects of the Moon are larger, the tidal effects of the Sun are not negligible. The high tides due to the Sun are determined by the solar day, and so have periods $T_s/2$ and T_s.

10.5 Tides on Lakes and Seas

Fermi briefly considers the problem of tides on finite bodies of water. Any finite body of water has certain natural frequencies. Let us consider a wide canal of length L and average depth h. In the limit $kh << 1$, the dispersion relation Eq. (9.34) reduces to $\omega = k\sqrt{gh}$, so $v_p = v_g = \sqrt{gh}$. For a wave of very large wavelength, we may estimate the tidal period by setting the velocity of the wave equal to L/T, which gives

$$T \sim \frac{L}{\sqrt{gh}}. \tag{10.30}$$

As an example, the Suez Canal has $L = 193\,\text{km}$, and $h = 24\,\text{m}$. Using Eq. (10.30) to estimate the period gives $T \sim 3.5\,\text{h}$.

Fermi also gives a brief treatment of a forced simple harmonic oscillator. The application to the problem at hand is to regard the natural frequency of the oscillator as the lowest frequency mode of the body of water under consideration. This is acted upon by an external force. Fermi notes that if the natural frequency of the oscillator is greater than the frequency of the force, the displacement of the oscillator and the force are in phase, while they are out of phase if the natural frequency of the oscillator is smaller than the force frequency. For the case of the Suez Canal, the passage of a vessel through the canal takes roughly 12 hours, so if passing vessels are treated as the external force on the canal, this is a case where the natural frequency of the canal is greater than the frequency of the applied force, so the applied force and natural motion of the canal are in phase. This is presumably not dangerous unless vessel passage time were to be reduced and approach the period of the canal.

(13)

I°	< 2.5 mm/sec^2	Instrumental
II°	2.5 – 5.0	only by few scattered persons
III	5 – 10	small percentage
IV	10 – 25	
V	25 – 50	felt by everybody
VI	50 – 100	
VII	100 – 250	
VIII	250 – 500	
IX	500 – 1000	
X	1000 – 2500	
XI	2500 – 5000	
XII	> 5000	

Origin San Francisco 4/18/1906

San Andreas 3 ÷ 6 m horiz

.5 ÷ 1 m vertical

10^8 dynes/cm^2

$$435 \text{ Km} \times 20 \text{ Km} \times \frac{4 \text{ m}}{2} = 1.7 \times 10^{24} \text{ erg}$$

The Mercalli scale of earthquake severity

CHAPTER 11

General Properties of Earthquakes

Fermi lists the Mercalli scale of earthquake intensity and estimates the energy released in the 1906 San Francisco earthquake. The Mercalli scale is a qualitative scale that classifies earthquakes by their local effects. For example, class I is a weak earthquake not felt by humans at the scene, while class V is moderate and felt by nearly everyone near the scene; class VIII is so severe it causes ordinary buildings to collapse.

"Great earthquakes" are ones that release enormous amounts of energy. Part of that energy is radiated as seismic waves, the properties of which are discussed in Chap. (12). The remainder of it goes into local heat and destruction. The latter is usually called "plastic deformation." The total energy released in the San Francisco earthquake of 1906 is generally estimated to be in the range of 10^{23-24} erg. The Earth Alabama website is helpful in grasping the significance of such a huge amount of energy (EarthAlabama 2017). It is noted there that the total energy released in the San Francisco earthquake was at least a thousand times larger than that released in the nuclear weapon detonated over Hiroshima in 1945.

A method of estimating the energy released in great earthquakes which has some overlap with Fermi's method, makes use of a quantity known as the "seismic moment" (Kanamori 1977). It is denoted as M_0 and defined as follows:

$$M_0 = \mu A \bar{D}, \tag{11.1}$$

where μ is the shear coefficient of the rock near the fault, A is the area of the fault that moves during the earthquake, and $\bar{D}$ is the average distance the area A moves during the earthquake. The product $A\bar{D}$ is a volume, while μ has the dimension of pressure, i.e., force/area, so M_0 has the dimension of energy.

Although it has the right units, M_0 is not the energy released in the earthquake. That can be seen by first considering the work done as the area A moves. Roughly speaking, work is force times distance, but account must be taken of the fact that the force moving the area A in the direction of its displacement is decreasing as the displacement increases. This produces a factor of 1/2 in the energy formula, similar to the 1/2 that appears in the energy of a charged capacitor.

The force in the direction of the displacement at any stage of the earthquake can be written as σA, where σ is the shear stress at the given displacement. If the net decrease in the stress is $\Delta\sigma$, then equating the work done to the energy released, we have

$$W_0 = \frac{1}{2}\Delta\sigma A\bar{D} = (\frac{\Delta\sigma}{2\mu})M_0. \tag{11.2}$$

The quantity $\Delta\sigma$ is known to be $\sim (2-6) \times 10^7$ dyne/cm^2, and for rock in the Earth's crust, $\mu \sim (3-6) \times 10^{11}$ dyne/cm^2

(Kanamori 1977). The ratio $2\mu/\Delta\sigma$ is dimensionless and is $\sim 2 \times 10^4$, so W_0 becomes

$$W_0 \sim \frac{M_0}{2 \times 10^4}. \quad (11.3)$$

Using a number of different techniques, the value of the seismic moment for the San Francisco earthquake has been determined to be

$$(M_0)_{SF} \sim 10^{28}\,\text{erg}, \quad (11.4)$$

which gives, using Eq. (11.2),

$$(W_0)_{SF} \sim 5 \times 10^{23}\,\text{erg}. \quad (11.5)$$

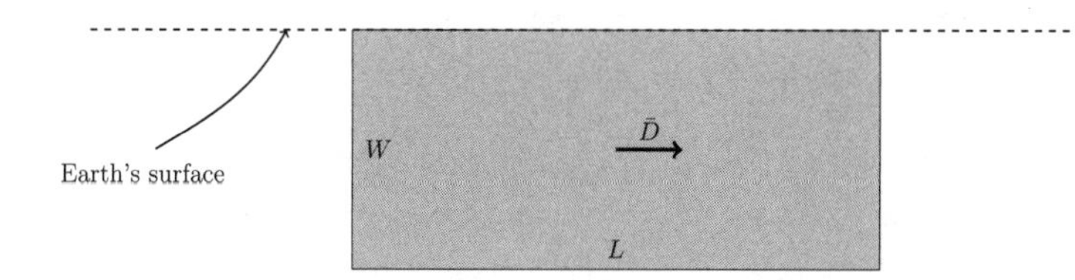

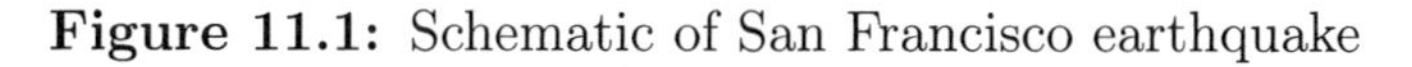

Figure 11.1: Schematic of San Francisco earthquake

In estimating the energy released in the San Francisco earthquake, Fermi used the first expression in Eq. (11.2). For this earthquake, the area of the fault involved in the earthquake can be approximated by a rectangle of area $A = LW$, where L is the horizontal dimension of A, and W is the vertical dimension. The standard values, as used by Fermi for L, W, and $\bar{D}$, were $L = 435\,\text{km}$, $W = 20\,\text{km}$, and $\bar{D} = 4\,\text{m}$. Fig. (11.1) shows schematically the geometry of the earthquake. Using these figures, the volume combination $A\bar{D}$ turns out to be $3.48 \times 10^{16}\,\text{cm}^3$. For the formula Fermi employs, he has used a figure of $10\,\text{dyne/cm}^2$ for $\Delta\sigma$, which leads to his result

$$(W_0)_{SF-Fermi} \sim 1.74 \times 10^{24}\,\text{ergs}, \quad (11.6)$$

approximately three times the value given in Eq. (11.5). Given the uncertainties in the San Francisco earthquake parameters and the fact that Fermi made his estimate in the early 1940s, it is remarkable that his value is within a reasonable factor of the generally accepted result.

(14)

Seismographs

horizontal

Damping

Theory of elastic waves (p 448)

456

Km	P	S - P
200	31s	23s
300	44	34
400	57	45
500	1m 10s	56
1000	2:15	1:47
2000	4:18	3:20
3000	5:58	3:41

(Table p. 501)

min

20

10

transv

S

long

P

10000 Km

~~First group P~~

$1.78 = \frac{v_l}{v_t}$

Seismographs and earthquake classification

CHAPTER 12

Seismic Waves and Seismology

12.1 Elementary Seismographs

The basic idea behind the workings of an elementary seismograph, as described, e.g., in Richter's text on seismology (Richter 1958), is the analysis of the changes induced by seismic waves in the behavior of a simple oscillator. As an example let the oscillator in question be an ordinary simple pendulum of length l and mass m. Newton's law, in the absence of any vibrational motion of the Earth, says that for small oscillations

$$m\ddot{x} = -\frac{mg}{l}x, \tag{12.1}$$

where x is the deviation from the vertical. When no external disturbances are present the pendulum oscillates at a natural frequency $\omega_0 = \sqrt{g/l}$.

Consider now the case where the pendulum's support is attached to the Earth's surface, which is itself oscillating at a frequency Ω because of seismic waves. The restoring force on the pendulum is still given by $-mgx/l$, but the acceleration of the mass must now also take into account the acceleration of the pendulum's support. Let X be the coordinate of that support relative to a fixed origin. The total acceleration of the mass is $\ddot{X} + \ddot{x}$, and Newton's law now reads

$$m(\ddot{X} + \ddot{x}) = -\frac{mg}{l}x = -m\omega_0^2 x, \tag{12.2}$$

which can be rewritten as

$$(\ddot{x} + \omega_0^2 x) = -\ddot{X}. \tag{12.3}$$

Setting $x(t) = x_0(t) + x_1(t)$,where $x_0(t)$ is any solution of Eq. (12.1), and $x_1(t)$ is a function that oscillates at frequency Ω, Eq. (12.3) becomes

$$\ddot{x}_1 + \omega_0^2 x_1 = -\Omega^2 X(t). \tag{12.4}$$

Solving for $x_1(t)$, we have

$$x_1(t) = \frac{\Omega^2}{\Omega^2 - \omega_0^2} X(t). \tag{12.5}$$

Now $x(t)$, the deviation from the vertical of the pendulum, is expressed as a superposition of an oscillation at the natural frequency of the pendulum and an oscillation at the frequency of the seismic wave. Disentangling the oscillation at frequency Ω is the goal of the device. To facilitate this analysis, damping of the oscillation at the natural frequency of the pendulum was introduced in early seismographs. Seismographs of the twenty-first century are much more sophisticated but make use of these same principles.

12.2 Seismic Waves

Having already examined the effects of earthquakes and seen a very simple example of how the oscillations they generate are measured, we now turn to the subject of how those waves are produced and how they propagate, taking note of the fact that there is more than one type of seismic wave. Prior to doing so, we review some basic concepts in the theory of elasticity, a subject often neglected in a common physics curriculum (Landau and Lifshitz 1986; Rawlinson). There are two key concepts. The first is that of strain, as encoded in a strain tensor that describes the deformations a body undergoes. The second is that of stress, with the body's stress tensor being the indicator of the forces leading to the deformations.

12.3 Strain and the Strain Tensor

When forces act on a solid, a small volume centered at $\boldsymbol{x}$ in a solid body moves to a new location

$$\boldsymbol{x}' = \boldsymbol{x} + \boldsymbol{u}(\boldsymbol{x}).$$

The fundamental quantity characterizing strain is not the displacement $\boldsymbol{u}(\boldsymbol{x})$, but the *strain*, a dimensionless measure of relative displacement.

Examples As an example, consider a bar of original length l, which is extended to length l' as shown in Fig. (12.1). If x is the coordinate in the undeformed bar, and x' is the corresponding coordinate in the deformed bar, we have

$$x' = \frac{l'}{l}x.$$

Figure 12.1: Elongation strain

The displacement in the x direction is

$$u_x(x) = (x' - x) = \frac{l' - l}{l}x.$$

The strain in this case is equal to

$$\frac{\partial u_x}{\partial x} = (\frac{l' - l}{l}). \tag{12.6}$$

We will consider only cases where the strain is small, so quadratic terms can be ignored. For our deformed bar, small strain means

$$|\frac{l' - l}{l}| << 1.$$

Another example of strain is shown in Fig. (12.2). Here the displacement in the x direction depends on y :

$$x' = x + \frac{\epsilon}{h}y, \ \ u_x(y) = x' - x = \frac{\epsilon}{h}y.$$

The strain in this case is

$$\frac{1}{2}\frac{\partial u_x}{\partial y} = \frac{\epsilon}{2h}. \tag{12.7}$$

The reason for the factor of 2 in the denominator of Eq. (12.7) will become clear below. (See Eq. (12.9).)

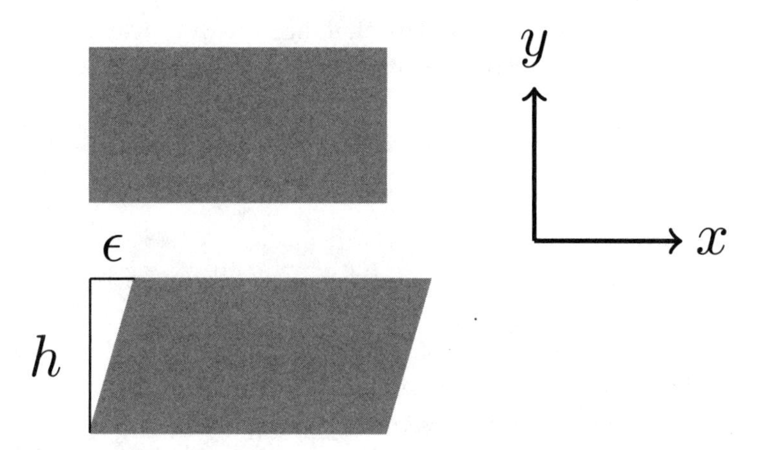

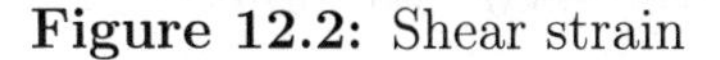

Figure 12.2: Shear strain

General Case Treating a general case, consider two points that are very close together in a solid body. Under deformation their separation $ds = \sqrt{dx_i dx_i}$ becomes

$$ds' = \sqrt{(dx_i + du_i)(dx_i + du_i)}.$$

Setting

$$du_i = \frac{\partial u_i}{\partial x_j} dx_j$$

in the above equation, we may write

$$ds' = \sqrt{ds^2 + 2u_{ij} dx_i dx_j},$$

where u_{ij} is a tensor, known as the strain tensor. It is easy to see that it equals

$$u_{ij} = \frac{1}{2}\left(\frac{\partial u_i}{\partial x_j} + \frac{\partial u_j}{\partial x_i} + \frac{\partial u_k}{\partial x_i}\frac{\partial u_k}{\partial x_j}\right). \qquad (12.8)$$

We will consider only cases of small strain,

$$\left|\frac{\partial u_i}{\partial x_j}\right| << 1.$$

Then the terms in Eq. (12.9) that are quadratic in partial derivatives of the displacement can be ignored and the strain tensor becomes

$$u_{ij} = \frac{1}{2}(\frac{\partial u_i}{\partial x_j} + \frac{\partial u_j}{\partial x_i}). \tag{12.9}$$

Volume Change Since strain changes distances between neighboring points, we also expect it to change small volumes. To see how, note that since u_{ij} is a symmetrical tensor, coordinate axes can be chosen at any point so that only its diagonal elements are nonzero. In such a coordinate system, a small volume dV in the undeformed solid transforms to dV' in the strained solid. We have

$$\begin{aligned} dV &= dx_1 dx_2 dx_3 \to dV' \\ &= dx_1(1 + u_{11})dx_2(1 + u_{22})dx_3(1 + u_{33}). \end{aligned} \tag{12.10}$$

Keeping only linear terms in strain tensor components,

$$dV' = dV(1 + u_{11} + u_{22} + u_{33}) = dV(1 + \mathrm{tr}(u))). \tag{12.11}$$

Due to the invariance of the trace, the last equality in Eq. (12.11) is true in any coordinate system, whether or not the strain tensor is diagonal.

12.4 Stress and the Stress Tensor

The total force on a volume V of a solid is the sum of the forces acting on the individual atoms in the volume. Treating the solid as a continuum, the total force is expressed as the integral over V of the force/volume, $\boldsymbol{f}(\boldsymbol{x})$. There are two types of terms in $\boldsymbol{f}(\boldsymbol{x})$, body forces and surface forces. Gravity is the most important body force.[1] Its contribution to $\boldsymbol{f}(\boldsymbol{r})$ is $\rho(\boldsymbol{x})\boldsymbol{g}(\boldsymbol{x})$, where $\boldsymbol{g}(\boldsymbol{x})$ is the local acceleration due to gravity. In Sec. (2.3), Eqs. (2.34)–(2.36) amounted to a calculation of $\boldsymbol{g}(\boldsymbol{x})$ for a spherical Earth.

The remaining term in the force/volume is the force on the material in V due to the material outside V. Using the fact that interatom forces are short in range, in the continuum this can be represented as a surface integral over the surface of V. Let us break $\boldsymbol{f}$ into terms arising from body and surface forces, $\boldsymbol{f} = \boldsymbol{f}_b + \boldsymbol{f}_s$. Doing likewise for the total force on the volume V, we can write $\boldsymbol{F}(V) = \boldsymbol{F}_b(V) + \boldsymbol{F}_s(V)$. The total body force is

$$\boldsymbol{F}_b(V) = \int_V \boldsymbol{f}_b(\boldsymbol{x}) d^3\boldsymbol{x}. \tag{12.12}$$

Likewise, the total surface force is

$$\boldsymbol{F}_s(V) = \int_V \boldsymbol{f}_s(\boldsymbol{x}) d^3\boldsymbol{x}. \tag{12.13}$$

To see how the volume integral in Eq. (12.13) can be transformed into a surface integral, it is easiest to work with specific components,

$$(F_s)_i = \int_V (f_s)_i d^3x. \tag{12.14}$$

By the generalization of Gauss's theorem, transformation of the volume integral to a surface integral is only possible if the integrand is a gradient,

$$(f_s)_i = \frac{\partial \sigma_{ij}}{\partial x_j}. \tag{12.15}$$

Using this form for $(f_s)_i$ in Eq. (12.14), we have

$$(F_s)_i = \int_V \frac{\partial \sigma_{ij}}{\partial x_j} d^3x = \int_S \sigma_{ij} dS_j, \tag{12.16}$$

[1]For rotating bodies, the centrifugal force is also a body force.

In Eq. (12.16), dS_j is the jth component of the element of surface $d\boldsymbol{S} = \hat{\boldsymbol{n}} dS$ where $\hat{\boldsymbol{n}}$ is the outward unit normal. The tensor σ_{ij} is known as the stress tensor. It is symmetric under interchange of indices, $\sigma_{ij} = \sigma_{ji}$, and has the same dimensions as pressure: force/area. To visualize its action, suppose a surface element of the body in question has its normal in the j direction. Regarding i as the "force index" and j as the "area index," σ_{ij} is the force/area in the ith direction acting on an element of the surface whose normal is in the jth direction.

Gravity and Stress In the next section, we will introduce Hooke's law for solids and then go on in subsequent sections to treat seismic waves. Body forces will play no role in this discussion. The reader may well wonder, what happened to body forces? Body forces do affect the stress tensor. They cause an equilibrium or zeroth-order stress. Hooke's law and seismic waves involve small oscillations around this state of stress. Writing the stress as the sum of zeroth and first-order parts, we have

$$\sigma_{ij} = \sigma_{ij}^{(0)} + \sigma_{ij}^{(1)}, \tag{12.17}$$

the zeroth-order term can be written

$$\sigma_{ij}^{(0)} = -\delta_{ij} p(r), \tag{12.18}$$

where the pressure $p(r)$ is the solution to Eq. (2.33). As discussed in Sec. (2.3), $p(r)$ depends on the entire mass inside radius r. So only for those displacements which affect the mass inside radius r does the body force due to gravity come into the equations of motion. Ordinary seismic waves are unaffected, but for oscillations of the Earth as a whole, the body force from gravity does come into the equations of motion. This point is discussed further in Sec. (13.3).

In the next sections on Hooke's law and seismic waves, it is to be understood that the part of the stress tensor being discussed is $\sigma_{ij}^{(1)}$, and the superscript will be dropped.

12.5 Hooke's Law

In the case of small deformations, strain is generally proportional to stress , a relation commonly referred to as Hooke's law. Its most familiar example is the harmonic oscillator, where the force on the oscillator is proportional to the displacement from equilibrium of the oscillator coordinate. The general statement of the proportionality between stress and strain is

$$\sigma_{ij} = C_{ij;kl} u_{kl}, \tag{12.19}$$

where the $C_{ij;kl}$ are a set of coefficients that are independent of stress and strain but may depend on position, temperature, etc. We will follow Fermi in only treating the simpler case of a homogeneous deformation, one in which the components of the strain tensor remain constant throughout the body. In this case of Hooke's law, the relation between the stress and the strain tensor only requires only two constants, Young's modulus and Poisson's ratio. These are introduced in the next section.

12.6 Young's Modulus and Poisson's Ratio

Fig. (12.3) shows a solid block held rigid below and with no external forces on the sides but subject to a pressure p from above, so only $\sigma_{33} \neq 0$. Similarly only the diagonal elements of $u_{ij} \neq 0$. The force/area acting on the top is directed downward, so $\sigma_{33} = -p$. The block will respond to the applied pressure by becoming "squished" or compressed, mean-

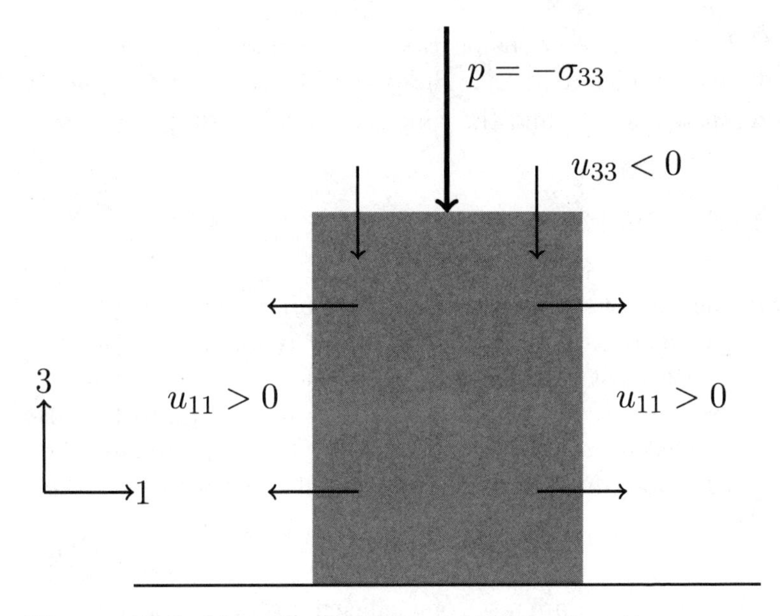

Figure 12.3: Young's modulus and Poisson's ratio

ing $u_{33} < 0$. Young's modulus E is the coefficient that relates stress to strain in the 3 direction:

$$\sigma_{33} = E u_{33}. \tag{12.20}$$

Strain is dimensionless, so from Eq. (12.20), we note that Young's modulus has the same units as stress or pressure, namely, force/area. The values of Young's modulus for solid materials are usually given in GPa, or 10^9 Pa. The order of magnitude of the Young's modulus of Earth materials $\sim$ 100 GPa, so for such materials even very small strains require pressures of thousands of atmospheres.

In addition to compression in the 3 direction, the block will generally expand in the perpendicular directions. For a homogeneous solid, the block expands equally in the two perpendicular directions, so $u_{11} = u_{22}$. The ratio of transverse to longitudinal strain is Poisson's ratio, σ. (In what follows, both Poisson's ratio and the stress tensor involve the Greek letter σ. They are distinguished by the stress tensor having two subscripts, while Poisson's ratio has none.) The defining equation is

$$u_{11} = u_{22} = -\sigma u_{33}. \tag{12.21}$$

The minus sign in this equation makes σ positive for all known solid materials. A solid with a value of $\sigma = 0.25$ is called a "Poisson solid."

12.7 The Stress-Strain Relation for a Solid

The components of the stress tensor of a material obeying Hooke's law will be proportional to the components of the strain tensor with the coefficients expressible in terms of Young's modulus and Poisson's ratio. Writing the strain tensor as the sum of of a trace and of a traceless part is an identity. We thus write

$$u_{ij} = \frac{1}{3}\delta_{ij}\mathrm{tr}(u) + \left(u_{ij} - \frac{1}{3}\delta_{ij}\mathrm{tr}(u)\right),$$

where $\mathrm{tr}(u) = u_{11} + u_{22} + u_{33}$. For a material obeying Hooke's law, having stress proportional to strain, we must have separately that the trace of the stress tensor is proportional to the trace of the strain tensor and the traceless part of the stress tensor is proportional to the traceless part of the strain tensor, so

$$\sigma_{ij} = K\delta_{ij}\mathrm{tr}(u) + 2\mu\left(u_{ij} - \frac{1}{3}\delta_{ij}\mathrm{tr}(u)\right), \tag{12.22}$$

where K and μ are constants. We saw earlier in Eq. (12.11) that a change in volume associated with strain is proportional to tr(u). The quantity K is therefore known as the bulk or compression modulus, while the second term, being traceless, represents a shear stress; hence μ is known as the shear modulus.

The two constants K and μ must be expressible in terms of Young's modulus and Poisson's ratio. To find the relationship, start with the trace of both sides of Eq. (12.22). This gives

$$\sigma_{11} + \sigma_{22} + \sigma_{33} = 3K(u_{11} + u_{22} + u_{33}). \tag{12.23}$$

Let us apply this equation to the situation of a compressed block, discussed earlier. In that situation, the only nonzero stress component is σ_{33}. On the right-hand side of Eq. (12.23), using Eq. (12.21), we can eliminate u_{11} and u_{22} in favor of u_{33}. Eq. (12.23) becomes

$$\sigma_{33} = 3K(-2\sigma + 1)u_{33}. \tag{12.24}$$

Using $\sigma_{33} = Eu_{33}$, we find

$$K = \frac{E}{3(1-2\sigma)}. \tag{12.25}$$

To determine μ, we write out the 33 component of Eq. (12.22). This is

$$\sigma_{33} = K(u_{11} + u_{22} + u_{33}) + 2\mu[\frac{2}{3}u_{33} - \frac{1}{3}(u_{11} + u_{22}]. \tag{12.26}$$

Again using the compressed block situation, we substitute for K, and using Eqs. (12.20) and (12.21), we see that

$$Eu_{33} = \left(\frac{E}{3} + \frac{4}{3}\mu(1+\sigma)\right)u_{33}, \tag{12.27}$$

or

$$\mu = \frac{E}{2(1+\sigma)}. \tag{12.28}$$

Since K and μ are both positive we must have

$$-1 \leq \sigma \leq 1/2.$$

In practice, since we saw earlier that $\sigma \geq 0$, our limit is

$$-0 \leq \sigma \leq 1/2.$$

Hooke's law in terms of the constants E and σ now reads

$$\sigma_{ij} = \frac{E}{3(1-2\sigma)}\delta_{ij}\mathrm{tr}(u) + \frac{E}{(1+\sigma)}(u_{ij} - \frac{1}{3}\delta_{ij}\mathrm{tr}(u)). \quad (12.29)$$

12.8 Newton's Law inside an Elastic Solid

Consider a volume V inside a solid. The surface of the volume is denoted as Σ. Newton's law for the mass of the solid inside V is, taking the ith component both of acceleration and of force,

$$\int_V \rho \frac{\partial^2 u_i}{\partial^2 t} dV = \int_\Sigma \sigma_{ij} dS_j, \quad (12.30)$$

where ρ is the mass density of the solid. Applying the divergence theorem, the right hand side of Eq. (12.30) becomes

$$\int_V dV \frac{\partial \sigma_{ij}}{\partial x_i},$$

and Newton's law now reads

$$\int_V dV \left(\rho \frac{\partial^2 u_i}{\partial^2 t} - \frac{\partial \sigma_{ij}}{\partial x_i}\right) = 0. \quad (12.31)$$

This equation must hold no matter what choice is made for the volume V, so the integrand must vanish, i.e.,

$$\rho \frac{\partial^2 u_i}{\partial^2 t} = \frac{\partial \sigma_{ij}}{\partial x_i}. \quad (12.32)$$

This is the generalization to the interior of a solid of the point particle version of Newton's laws, $\boldsymbol{F} = m\boldsymbol{a}$.

12.9 The Equation of Motion for Displacement

We want to turn Eq. (12.32) into a partial differential equation for the displacement $\boldsymbol{u}(\boldsymbol{x})$. Since u_{ij} is proportional to derivatives of position, and σ_{ij} is proportional to u_{ij}, the right-hand side of this partial differential equation will only involve second-order spatial derivatives of $\boldsymbol{u}$. In order for the right-hand side to also be a 3-vector, it must be a linear combination of the two possible 3-vectors, $\boldsymbol{\nabla}(\boldsymbol{\nabla} \cdot \boldsymbol{u})$, and $(\nabla^2)\boldsymbol{u}$. To discover the coefficients of the two, we return to Hooke's law as expressed in Eq. (12.22). Rearranging this equation slightly, we have

$$\frac{\partial \sigma_{ij}}{\partial x_j} = (\frac{\partial \mathrm{tr}(u)}{\partial x_i})(K - \frac{2}{3}\mu) + 2\mu \frac{\partial u_{ij}}{\partial x_j}. \tag{12.33}$$

Now

$$\mathrm{tr}(u) = u_{11} + u_{22} + u_{33} = \frac{\partial u_1}{\partial x_1} + \frac{\partial u_2}{\partial x_2} + \frac{\partial u_3}{\partial x_3} = (\boldsymbol{\nabla} \cdot \boldsymbol{u}).$$

Furthermore,

$$\frac{\partial u_{ij}}{\partial x_j} = \frac{1}{2}\frac{\partial}{\partial x_j}(\frac{\partial u_j}{\partial x_i} + \frac{\partial u_i}{\partial x_j}) = \frac{1}{2}(\boldsymbol{\nabla}_i(\boldsymbol{\nabla} \cdot \boldsymbol{u}) + (\nabla^2)u_i).$$

Returning to the derivative of the stress tensor, we have

$$\frac{\partial \sigma_{ij}}{\partial x_j} = (K + \frac{1}{3}\mu)\boldsymbol{\nabla}_i(\boldsymbol{\nabla} \cdot \boldsymbol{u}) + \mu(\nabla^2)u_i). \tag{12.34}$$

Substituting for K and μ from Eqs. (12.25) and (12.28), the final form of the equation of motion is

$$\rho \frac{\partial^2 \boldsymbol{u}}{\partial t^2} = \frac{E}{2(1 - 2\sigma)(1 + \sigma)} \boldsymbol{\nabla}(\boldsymbol{\nabla} \cdot \boldsymbol{u}) \tag{12.35}$$

$$+ \frac{E}{2(1 + \sigma)}(\nabla^2)\boldsymbol{u}.$$

This equation describes the waves that can propagate in an elastic solid.

12.10 Longitudinal and Transverse Waves

Eq. (12.35) is easy to solve for an infinite medium of fixed composition. An arbitrary solution can be expanded into plane waves of the form

$$\boldsymbol{U}(\boldsymbol{k},t)e^{i\boldsymbol{k}\cdot\boldsymbol{x}}. \tag{12.36}$$

Acting on this functional form, $\boldsymbol{\nabla}$ brings down a factor of $i\boldsymbol{k}$, reducing Eq. (12.35) to an equation for the amplitude $\boldsymbol{U}(\boldsymbol{k},t)$,

$$\rho(\frac{d^2\boldsymbol{U}}{dt^2}) = -\frac{E}{2(1+\sigma)}[\frac{1}{(1-2\sigma)}\boldsymbol{k}(\boldsymbol{k}\cdot\boldsymbol{U}) + (\boldsymbol{k}\cdot\boldsymbol{k})\boldsymbol{U}]. \tag{12.37}$$

We resolve the amplitude $\boldsymbol{U}$ into components parallel and perpendicular to $\boldsymbol{k}$, or longitudinal (L) and transverse (T) parts:

$$\boldsymbol{U} = \boldsymbol{U}_L + \boldsymbol{U}_T. \tag{12.38}$$

Our equation of motion now becomes

$$\rho(\frac{d^2\boldsymbol{U}_L}{dt^2} + \frac{d^2\boldsymbol{U}_T}{dt^2}) = -k^2\frac{E}{2(1+\sigma)}[\frac{2(1-\sigma)}{(1-2\sigma)}\boldsymbol{U}_L + \boldsymbol{U}_T]. \tag{12.39}$$

Every term in the equation can be resolved into longitudinal and transverse components. Taking the longitudinal case first, we have

$$\rho(\frac{d^2\boldsymbol{U}_L}{dt^2}) = -k^2\frac{E(1-\sigma)}{(1+\sigma)(1-2\sigma)}\boldsymbol{U}_L. \tag{12.40}$$

This is just a harmonic oscillator equation whose solution is

$$\boldsymbol{U}_L(\boldsymbol{k},t) = \boldsymbol{U}_L(\boldsymbol{k})e^{-i\omega_L t}, \tag{12.41}$$

where the frequency satisfies

$$\rho\omega_L^2 = \frac{E(1-\sigma)}{(1+\sigma)(1-2\sigma)}k^2. \tag{12.42}$$

To summarize, a longitudinal plane wave solution of Eq. (12.35) takes the form

$$\boldsymbol{U}_L(\boldsymbol{k})e^{i\boldsymbol{k}\cdot\boldsymbol{x}-i\omega_L t}, \tag{12.43}$$

where $\boldsymbol{k} \times \boldsymbol{U}_L = 0$. The velocity of the longitudinal waves, c_L, is given by

$$c_L^2 = (\frac{\omega_L}{k})^2 = \frac{E(1-\sigma)}{\rho(1+\sigma)(1-2\sigma)}. \tag{12.44}$$

Proceeding in a similar way, the transverse component of Eq. (12.39) satisfies

$$\rho(\frac{d^2\boldsymbol{U}_T}{dt^2}) = -k^2\frac{E}{2(1+\sigma)}(\boldsymbol{U}_T), \tag{12.45}$$

which is solved by

$$\boldsymbol{U}_T(\boldsymbol{k},t) = \boldsymbol{U}_T(\boldsymbol{k})e^{-i\omega_T t}, \tag{12.46}$$

with the frequency given by

$$\rho\omega_T^2 = \frac{E}{2(1+\sigma)}k^2. \tag{12.47}$$

A transverse plane wave solution of Eq. (12.35) is

$$\boldsymbol{U}_T(\boldsymbol{k})e^{i\boldsymbol{k}\cdot\boldsymbol{x}-i\omega_T t}, \tag{12.48}$$

where $\boldsymbol{k} \cdot \boldsymbol{U}_T = 0$. The velocity of transverse waves is c_T, where

$$c_T^2 = (\frac{\omega_L}{k})^2 = \frac{E}{2\rho(1+\sigma)}. \tag{12.49}$$

The ratio of transverse to longitudinal velocities is a dimensionless function of Poisson's ratio. From Eqs. (12.44) and (12.49), we have

$$(\frac{c_T}{c_L})^2 = 1 - \frac{1}{2(1-\sigma)}. \tag{12.50}$$

As we saw earlier, $-1 \le \sigma \le 1/2$, so longitudinal waves have higher velocity than transverse ones. This also means they arrive at detectors earlier. causing them to be known by seismologists as P or "primary waves." Transverse waves are accordingly known as S or "secondary waves." Reflecting this choice, the wave velocities are also often denoted by v_P and v_S rather than c_L and c_T. Their typical values in the Earth's continental crust are, respectively, of order 5 km/s and 3 km/s (Richter 1958). Using these values and Eq. (12.50), we find that Poisson's ratio in the Earth's continental crust is $\sigma \sim 0.22$, a conclusion consistent with the measurements of materials in that crust that generally show Poisson ratios lying in the range $0.20 - 0.30$.

The density of the continental crust is $\rho \sim 2.7\,\mathrm{g/cm^3}$. Using this value, $v_P = 5\,\mathrm{km/s}$, Eq. (12.44), and taking $\sigma = 0.25$, we find an average value for Young's modulus in the continental crust of $E \sim 59$ Gpa, about $1/4$ the value of Young's modulus for iron or steel, and otherwise typical of rocks found in the Earth's surface.

Studying deeper regions of the Earth, namely, moving beyond the crust into the mantle, we see that P and S wave velocities both increase. At a depth of around 3000 km, $v_P = 13.5\,\mathrm{km/s}$ and $v_S = 8\,\mathrm{km/s}$ (Richter 1958). Beyond the mantle, we enter the Earth's core, whose outer part is believed to be fluid. This is in agreement with the observation that S waves do not penetrate this region, as expected because liquids cannot support the shear stress needed for their existence. However, longitudinal (P) waves can and do pass through this region. As for the frequencies of seismic waves, they are generally in the $1 - 10\,\text{Hz}$ range, with corresponding wavelengths ranging from a few hundred meters to several thousand meters.

12.11 Snell's Law for Elastic Waves

Seismic waves, like optical waves, are refracted and reflected when they strike a boundary between two media. Their wave vectors are determined by a generalized form of Snell's law, familiar from optics. Suppose the boundary between the two media is the plane $x_3 = 0$, and the wave vectors of all waves are in the $1 - 3$ plane. Fig. (12.4) shows three such waves: (1) an incident wave at angle ϕ_a with respect to the normal; (2) a reflected wave also making an angle ϕ_a with respect to the normal; and (3) a transmitted wave making an angle ϕ_b with respect to the normal. In optics, Snell's law is usually expressed in terms of indices of refraction,

$$n_a \sin\phi_a = n_b \sin\phi_b. \tag{12.51}$$

It can equally well be written using the velocity of light propagation in the media a and b,

$$\frac{\sin\phi_a}{v_a} = \frac{\sin\phi_b}{v_b}, \tag{12.52}$$

where we use $v_a = c/n_a, v_b = c/n_b$. Fig. (12.4) shows a case where $\phi_b < \phi_a$, corresponding to $v_b < v_a$. This is an illustration of the familiar phrase in optics that "going from a fast medium to a slow medium, the transmitted wave is bent toward the normal."

Seismic waves also obey Snell's law, but in general seismic waves cannot be described by the situation of elementary op-

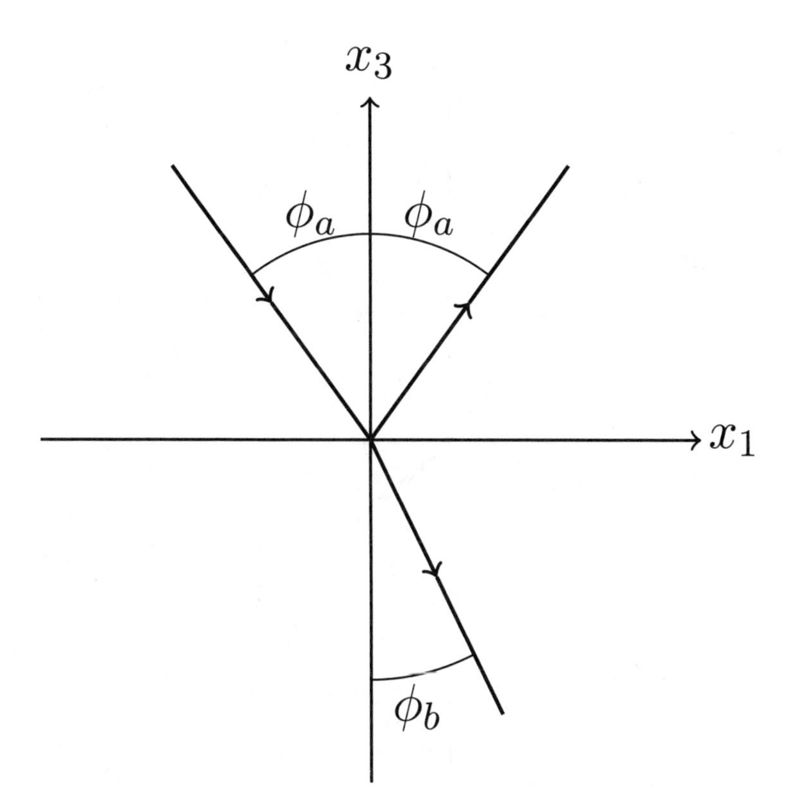

Figure 12.4: Three waves at a boundary

tics where a single reflected wave and a single transmitted wave are sufficient The main difference is that requiring the continuity of displacement and of the stress tensor leads to four boundary conditions at the interface. Given an incident wave of known amplitude, four waves must be present: reflected S and P waves, and transmitted S and P waves. In Fig. (12.5) we illustrate the situation for an incident P wave. The situation illustrated in the figure is for the case where $x_3 > 0$ is the "slow" medium, and $x_3 < 0$ is the "fast" medium. In going from slow medium to fast medium, waves are bent

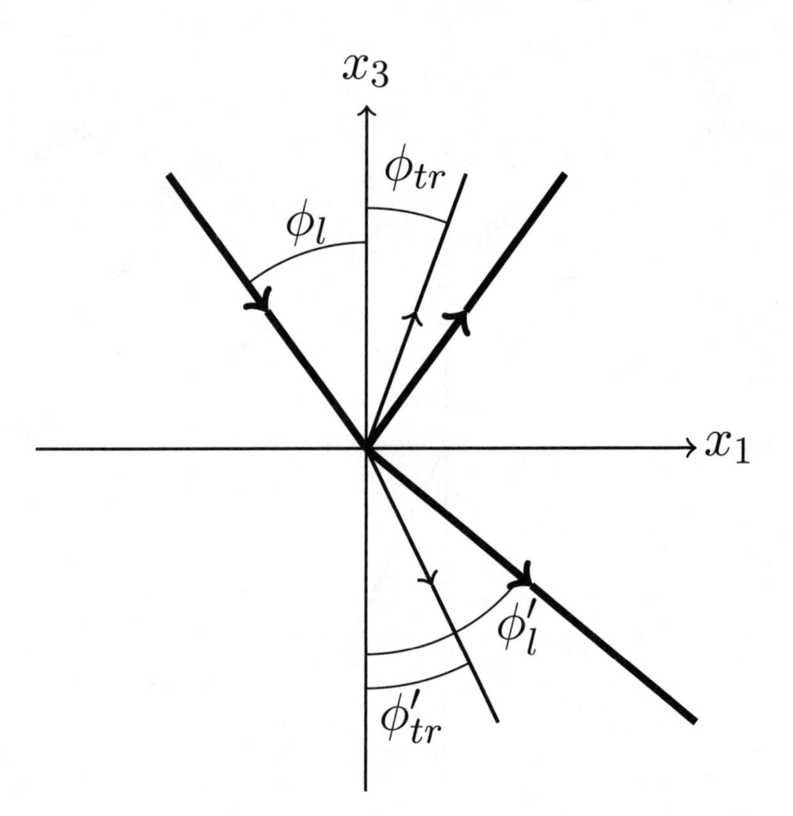

Figure 12.5: P wave incident on a boundary
Note: Thick lines are P waves; thin lines are S waves.

away from the normal. In each medium the longitudinal velocity is greater than the transverse velocity, so transverse waves are closer to the normal. If v_l, v_{tr} are the velocities of P and S waves in the upper medium, and v_l', v_{tr}' are for the lower medium, the statement of Snell's law for this situation is

$$\frac{\sin\phi_l}{v_l} = \frac{\sin\phi_l}{v_l} = \frac{\sin\phi_{tr}}{v_{tr}} = \frac{\sin\phi_l'}{v_l'} = \frac{\sin\phi_{tr}'}{v_{tr}'}, \tag{12.53}$$

where the first term is written twice to indicate that incident and reflected P waves have the same angle ϕ_l. These conditions follow simply, as in optics, from the constancy of the frequency and of the tangential component of the wave vector in the two media. The situation depicted in Fig. (12.5) occurs in a general way in the Earth, with the crust as the upper medium, and the mantle as the lower medium. The actual situation in the Earth is more complex due to a break in rock composition known as the Mohorovic discontinuity. Lying at the boundary between crust and mantle, this discontinuity is up to 500 m thick.

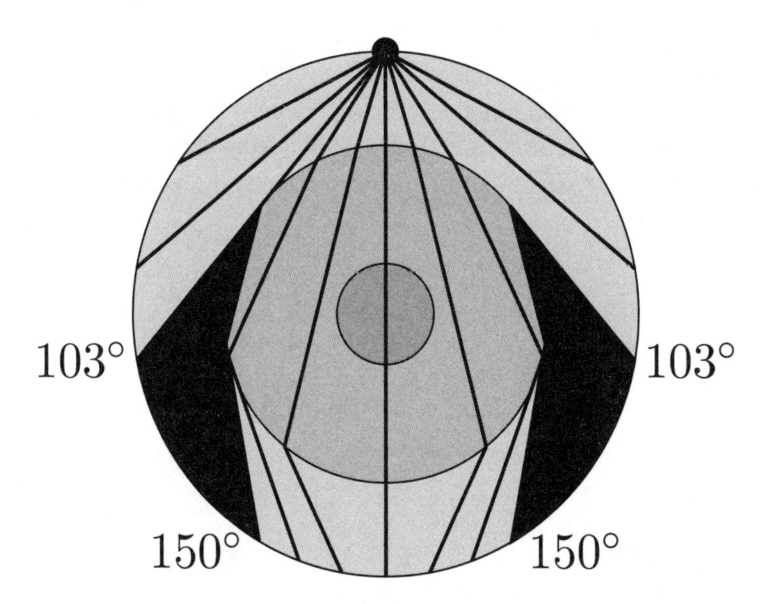

Figure 12.6: Paths for P waves

Fig. (12.6) illustrates the pattern of P seismic waves following an earthquake. The focus of the earthquake is at the top of the graph, and the straight black lines are possible paths of P seismic waves.[2] The large circles represent man-

[2]Actually, the lines have slight curvature, but they are shown here as

tle, outer core, and inner core, respectively. Taking the focus of the earthquake to be at 0°, the blacked out regions between 103° and 150° are angles between which no P waves can be received from the earthquake. The explanation is simply Snell's law physics. The mantle is the "fast" medium and the core is the "slow" medium. The ray that just grazes the core arrives at the surface of the Earth at 103°. This ray can also be refracted into the core and after refraction reaches the Earth's surface at 150°. A similar phenomenon would occur with light if the mantle were replaced by air (fast medium), and the core (slow medium) by glass.

Another striking phenomenon on the side of the Earth away from the earthquake is that S waves are not detected at any angles greater than 103°. This is very strong evidence that the outer core is liquid, where transverse waves cannot propagate.

---

straight for simplicity.

(15)

Rayleigh-waves

$$u = \left[A e^{-z\sqrt{f^2 - p^2/a^2} + igx + iky} + B e^{-z\sqrt{f^2 - p^2/b^2} + igx + iky} \right] e^{ipt}$$

normal & tangential stress = 0 on the surface

horizontal comp

vertical component

Other wave types with sheets have dispersion (Q-waves

group velocity

P first waves
S second waves
L long waves
C Tail waves

P			
0	5.6	3.5	
	5.9	3.75	3.4
60	8.0	4.3	
1200	12.5	6.75	
1790	12.75	7.25	
2450	13.25	7.25	6.39
2900	13.0	7.25	
[illegible]0	8.5		9.60
6370	11		

Surface seismic waves

CHAPTER 13

Surface Seismic Waves and Oscillations of the Earth

13.1 Surface Waves

There are waves concentrated near the surface in addition to the S and P waves that travel through the Earth. These, generally denoted as L waves, are of two types, sometimes known as Rayleigh (LR) and Love (LQ) waves after Lord Rayleigh and Augustus Love, who deduced the waves' existence. As revealed by their analyses, the velocity of these surface waves is smaller then that of either S or P waves. Consequently surface waves are the last to arrive at a seismograph from a distant disturbance. But despite their slower velocities, surface waves from an earthquake are very destructive because their intensity falls off only as the inverse distance from the disturbance, while S and P waves lose intensity as the inverse square of distance. This is illustrated in Fig. (13.1) where it is seen that although L waves arrive at a recording station

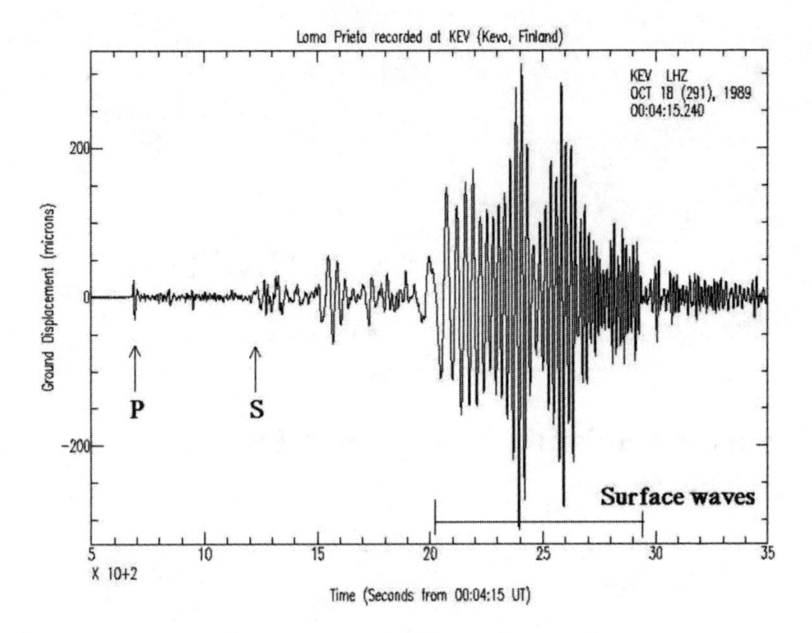

Figure 13.1: Seismic record of 1989 Loma Prieta earthquake
Source: USGS

later than S and P waves, their amplitudes are considerably larger.

Rayleigh and Love waves can be distinguished in a seismograph that can detect all three directions of oscillation. Taking the local vertical direction to be the z direction, suppose the surface wave is traveling in the x direction. Then a Rayleigh wave has displacements in the z and x directions, while the displacement of a Love wave is entirely in the y direction. The treatment of Love waves is somewhat intricate and will not be covered here. However, many interesting properties of Rayleigh waves are easily derived; an elementary treatment is given in Appendix C.

13.2 Oscillations of the Earth

In previous sections, we have considered the Earth as a medium through which seismic waves travel, much as light can travel through a medium with a variable index of refraction. In this section, we consider the oscillations of the Earth as a whole, a problem first considered in the nineteenth century by S. D. Poisson, Lord Rayleigh, and Horace Lamb, among others. The equation of motion for the various seismic waves applies here as well.

We will confine the discussion here to the simplest possible mode of oscillation, namely, the "breather" mode in which the Earth undergoes only radial oscillations. Fermi does not discuss this in his lectures, but it provides a possible example of a "Fermi question," the type of question that Fermi constantly posed to his colleagues and students. He might have phrased it as "Given the radius of the Earth and making an estimate of the Earth's average elastic properties, what is the order of magnitude of the fundamental period of the radially symmetric mode of oscillation?" The next section answers that question, and the following section gives more details on the breather mode of the Earth.

13.3 Breather Mode

Order of Magnitude of the Period Since the the displacement is purely radial in a breather mode it locally looks like a longitudinal or P wave. The displacement $\boldsymbol{u}$ is a vector, but assuming the equation of motion for $\boldsymbol{u}$ can be reduced to a scalar wave equation for a scalar function ψ, it should satisfy the wave equation

$$\frac{1}{c_L^2}\frac{\partial^2 \psi}{\partial t^2} = \nabla^2 \psi, \tag{13.1}$$

where c_L^2 is the squared velocity of longitudinal seismic waves, defined in Eq. (12.44). Assuming a frequency ω_b, Eq. (13.1) can be written

$$(\nabla^2 + k_b^2)\Psi = 0, \tag{13.2}$$

where $\omega_b^2 = (k_b c_L)^2$. The only spherically symmetric solution of of Eq. (13.2) which is finite at the origin is

$$\frac{\sin(k_b r)}{k_b r},$$

so ψ takes the form

$$\psi = A_b \frac{\sin(k_b r)}{k_b r}, \tag{13.3}$$

for some constant A_b. The value of k_b will be determined by a boundary condition at $r = a$. Once a definite k_b is in hand, the period is determined. We have

$$\omega_b = \frac{2\pi}{T_b} = k_b c_L, \quad \text{or} \quad T_b = \frac{2\pi}{k_b c_L}. \tag{13.4}$$

Since we have not specifically identified ψ, it is not obvious what boundary condition to apply. A reasonable guess for the order of magnitude of k_b is obtained by requiring ψ to vanish at $r = a$. Requiring $\psi(a) = 0$ gives $k_b a = \pi$. Setting $k_b = \pi/a$, our formula for the period becomes

$$T_b = \frac{2a}{c_L}. \tag{13.5}$$

Since our goal here is to obtain an order of magnitude estimate, we evaluate T_b from Eq. (13.5) using a range of values of

c_L; 5 km/s, 10 km/s, and 15 km/s. Taking $a = 6371$ km, these yield $T_b \sim 42$ min, 21 min, and 14 min, respectively. The answer to Fermi's hypothetical question on the rough magnitude of T_b is then "a few tens of minutes." In the next section, we will analyze T_b more precisely.

The Breather Period for a Homogeneous Earth In this section, we fill in the details and derive the breather period for a homogeneous Earth with constant elastic properties. Since the displacement is purely radial, it satisfies $\boldsymbol{\nabla} \times \boldsymbol{u} = 0$. Therefore

$$\boldsymbol{\nabla} \times (\boldsymbol{\nabla} \times \boldsymbol{u}) = \boldsymbol{\nabla}(\boldsymbol{\nabla} \cdot \boldsymbol{u}) - \nabla^2 \boldsymbol{u} = 0. \tag{13.6}$$

As a result, both terms on the right-hand side of the equation of motion for $\boldsymbol{u}$ (Eq. (12.35)) are of the same form. Assuming a harmonic time dependence with frequency ω_b, Eq. (12.35) becomes

$$-\omega_b^2 \boldsymbol{u} = c_L^2 \boldsymbol{\nabla}(\boldsymbol{\nabla} \cdot \boldsymbol{u}) = 0, \tag{13.7}$$

where c_L^2 is defined as before by Eq. (12.44). Setting $\omega_b^2 = (k_b c_L)^2$, Eq. (13.7) becomes an equation for $\boldsymbol{u}$, given $\boldsymbol{\nabla} \cdot \boldsymbol{u}$,

$$\boldsymbol{u} = -\frac{1}{k_b^2} \boldsymbol{\nabla}(\boldsymbol{\nabla} \cdot \boldsymbol{u}). \tag{13.8}$$

To find $\boldsymbol{\nabla} \cdot \boldsymbol{u}$, we take the divergence of both sides of Eq. (13.7) and obtain

$$(\nabla^2 + k_b^2)(\boldsymbol{\nabla} \cdot \boldsymbol{u}) = 0. \tag{13.9}$$

Comparing with Eq. (13.2), the scalar function ψ used in the previous section can be identified as $\boldsymbol{\nabla} \cdot \boldsymbol{u}$. Again requiring that the solution for $\boldsymbol{\nabla} \cdot \boldsymbol{u}$ be spherically symmetric and non-singular at $r = 0$, $\boldsymbol{\nabla} \cdot \boldsymbol{u}$ must take the form of Eq. (13.3):

$$\boldsymbol{\nabla} \cdot \boldsymbol{u} = A_b \frac{\sin(k_b r)}{k_b r}. \tag{13.10}$$

The period of the breather mode is determined by imposing a physical boundary condition at the Earth's surface. The pressure of the atmosphere at the Earth's surface is negligible compared to pressures that occur inside the Earth. The correct boundary condition is then that there should be vanishing stress exerted by the breather mode normal to the Earth's surface. For an Earth which is a homogeneous sphere, the only component of stress at the boundary is σ_{rr}, so the boundary condition is that $\sigma_{rr}(a) = 0$. The stress tensor is given in terms of the strain tensor by Eq. (12.29). Omitting some algebraic details, it is possible to use this equation to find $\sigma_{rr}(r)$ in terms of $\boldsymbol{\nabla} \cdot \boldsymbol{u}$. The formula is

$$\sigma_{rr}(r) = \frac{E}{(1+\sigma)(1-2\sigma)}[(1-\sigma)(\boldsymbol{\nabla} \cdot \boldsymbol{u}) + \frac{2(1-2\sigma)}{(k_b r)^2} r \frac{\partial}{\partial r}(\boldsymbol{\nabla} \cdot \boldsymbol{u})]. \tag{13.11}$$

Requiring that σ_{rr} vanish at $r = a$ gives

$$[(1-\sigma)(\boldsymbol{\nabla} \cdot \boldsymbol{u}) + \frac{2(1-2\sigma)}{(k_b r)^2} r \frac{\partial}{\partial r}(\boldsymbol{\nabla} \cdot \boldsymbol{u})]_{r=a} = 0. \tag{13.12}$$

From this equation, we see that the correct boundary condition requires the vanishing of a combination of the value and derivative of $\boldsymbol{\nabla} \cdot \boldsymbol{u}$. Setting $\boldsymbol{\nabla} \cdot \boldsymbol{u} = 0$ would only be correct for $2\sigma = 1$. Using Eq. (13.10) for $\boldsymbol{\nabla} \cdot \boldsymbol{u}$, it is easy to show that Eq. (13.12) is equivalent to (Stoneley 1961)

$$(k_b a) \cot(k_b a) = 1 - \frac{1}{2}(\frac{1-\sigma}{1-2\sigma})(k_b a)^2. \tag{13.13}$$

It is reasonable to assume the Earth is isotropic, which corresponds to a Poisson ratio $\sigma = 0.25$. For $\sigma = 0.25$, the smallest root of Eq. (13.13) is at $k_b a = 2.563$.[1] The corresponding pe-

[1] Note that this is a smaller value of $k_b a$ than the guessed value of π.

riod is given by

$$T_b = \frac{2\pi}{k_b a}(\frac{a}{c_L}) = \frac{2\pi}{2.563}(\frac{a}{c_L}). \tag{13.14}$$

The period of the breather mode has been observed (Ness, Harrison, and Slichter 1961) to be approximately 20.5 min, or 1230 s. We may use this value for T_b along with the known Earth radius of $a = 6371$ km to deduce a value of c_L. The speed of longitudinal waves is known to be different for different regions of the Earth, but since the mantle occupies the greatest volume, we might expect to obtain a value for c_L characteristic of the mantle. Using Eq. (13.14) to solve for c_L gives

$$c_L = \frac{2\pi a}{k_b a T_b} = \frac{2\pi(6371)}{2.563(1230)}\,\text{km/s} = 12.7\,\text{km/s}, \tag{13.15}$$

a reasonable value for c_L in the Earth's mantle.

While assuming the Earth has constant elastic properties is clearly an approximation, it would appear that an even more drastic approximation is the complete neglect of gravity inside the Earth. As we saw in Sec. (2.3), due to the inward pull of gravity, the pressure at the center of the Earth is millions of times larger than atmospheric pressure. Thus even at equilibrium, the deep interior of the Earth is in a state where both stress and strain are definitely nonzero. As realized by several physicists in the nineteenth century, the oscillations of the Earth should be thought of as small oscillations about a state of hydrostatic equilibrium between the inward force of gravity and the stress tensor of the Earth's materials. The case of an Earth-sized sphere with constant elastic properties was analyzed in detail by A. H. Love (Love 1911), among others. The general effect of gravity on the period of an Earth oscillation is to lower the period. However, due to its complete spherical symmetry, the breather frequency is unaffected, so the analysis given above is correct for an Earth with constant elastic properties even when gravity is taken into account.

The first observation of a normal mode of vibration of the Earth took place in the 1950s. By now many higher modes have been observed with nontrivial angular dependence (Stein and Wysession 2003). Just as plucking a violin string excites its normal modes, earthquakes excite the normal modes of the Earth. A huge effort has gone into explaining the data on the various modes of oscillation of the Earth in terms of its internal properties (Schubert 2009). While the breather mode discussed above is consistent with a homogeneous and isotropic Earth, the higher modes are not, and they require models which take detailed account of the variation in the Earth's parameters in its different regions.

(16)

Estimates of density in the interior are based on
a) Total Mass
b) Ellipticity of the meridian
c) Difference of the moments of inertia
d) Density at the surface
Assume hydrostatic equilibrium
Various analytical expressions

$$\rho = \rho_0 \left[1 - \beta \frac{r^2}{R^2}\right] \qquad \rho_0 = 10.1 \quad \beta = 0.764 \quad \text{(Roche)}$$

$$\rho = \rho_0 \left[1 - \beta_1 \frac{r^2}{R^2} + \beta_2 \frac{r^4}{R^4}\right] \qquad \rho_0 = 11.3 \quad \beta_1 = 1.04 \quad \beta_2 = 0.275$$

(Helmert)

Discontinuous laws
Gutenberg

				linear to	Constant	
Linear from	3.4 to	4.7	4.7	5	11	11
Depth	60 Km	1200	2450	2900	2900	6370

Radioactivity
Measurements of emanation (Ra & Th)
3.85 days 54 sec

Ac U 7.13×10^8	4.56×10^9 y	U
	1.31×10^{10} y	Th
	1690 y	Ra

Granit	2.7×10^{-12} Ra	2.00×10^{-5} Th
Basalt	1.4×10^{-12}	0.56×10^{-5}
Stone meteorites	0.7×10^{-12}	
Iron	0.05×10^{-12}	

	produces
U	2.5×10^{-8} cal/gr sec
Th	6.6×10^{-9}
K	3.9×10^{-12}

Granit	20 Km	2.6% K	6×10^{-6} U	2×10^{-5} Th	3.2×10^{20} cal/y
Basalt	40	1.28	1.8	0.6	2.6
Peridotit	1540	0.22	0.3	0.1	16.1
Iron free	700	0.14	0.2	0.06	4.3
					26.2×10^{20} cal/y

5.1×10^{18} × (10 Km) = 5.1×10^{24} cm^3 $\qquad \frac{10^{13}}{10^{25}}$

Radioactivity of rocks in Earth's crust and mantle

CHAPTER 14

Radioactivity and the Earth's Interior

14.1 Historical Note

The discussion in Fermi's notes of the density in the Earth's interior is very short, scarcely half a page. However, the brief summary notes of the Sixth Washington Conference on Theoretical Physics show this to be a subject that interested Fermi greatly at the time these notes were written and one in which he was regarded as an expert.

The yearly three-day Washington Conferences on Theoretical Physics began in 1935. They were the brainchild of the physicist and, later, cosmologist Russian refugee George Gamow, then a recent arrival in the United States. Aided in their organization by fellow refugee Hungarian Edward Teller and the Carnegie Institution's Merle Tuve, they loosely modeled the meetings on the annual gatherings hosted by the eminent Danish physicist Niels Bohr in Copenhagen during which a small group of physicists gathered for informal discussions not marked by an official record of the proceedings.

One way in which Washington differed from Copenhagen was that a topic for discussion was chosen each year and the group attending was constituted with this in mind. The Washington meetings were a great success from the start, in part because they provided an informal atmosphere for American physicists to become acquainted with recent emigres such as Gamow and Teller.

The 1938 conference, with topic of "Stellar Energy and Nuclear Processes," is notable for drawing the leading nuclear physicist Hans Bethe into applying his expertise to astrophysics. The 1939 conference, whose topic was "Low Temperature Physics and Superconductivity," is best remembered for the organizers' decision to abandon the planned schedule and have the first two talks be on nuclear fission, a discovery that had been announced only weeks earlier. The impact of the talks was magnified by the universal respect for the first two speakers, Niels Bohr and Enrico Fermi.

14.2 The Sixth Washington Conference on Theoretical Physics

The chosen topic for 1940 was "The Interior of the Earth." Though, as stated earlier, proceedings were not published, a brief summary of the discussions was issued, giving us some idea of what transpired. The first day's discussions, centering on the physical state and composition of the Earth's interior, were introduced and led by Fermi. This was a mark of both his interest and expertise in the subject. Fermi began by asserting that if electrons could be treated as moving independently, the pressure-volume relation could be dealt with by statistical reasoning, but of a very different sort than is used at normal pressures, densities, and temperatures. It was known that the center of the Earth had a mass density of approximately $12\,\mathrm{g/cm^3}$, and was composed predominantly of iron. Fermi was interested in the question of whether the pressure there could be explained by assuming that the iron atoms were completely stripped of their electrons, and that the electrons formed a free gas obeying what is now called "Fermi-Dirac" statistics. If the temperature is very low, this is a completely quantum mechanical regime. For particles of mass m and number density ρ, the pressure of an ideal Fermi gas is (Landau and Lifshitz 1980)

$$P = \frac{\pi(9\pi)^{1/3}}{5m_e}\hbar^2\rho^{5/3}, \tag{14.1}$$

where $\hbar$ is Planck's constant divided by 2π. Using a mass density for the Earth of $12\,\mathrm{g/cm^3}$, and assuming the composition is 100% iron, the density of electrons is

$$\rho = 3.37 \times 10^{24}\,\mathrm{electrons/cm^3}.$$

For electrons stripped from the iron nuclei, Fermi's suggestion was to assume they form an ideal Fermi gas, in which case Eq. (14.1) can be used to compute the pressure. Using the known values of $\hbar$ and the electron mass, this gives

$$P_{earth-center} = 1.7 \times 10^8 P_{atmosphere}.$$

This is much higher than the estimate found by Fermi as described in Sec. (2.3). Fermi correctly concluded that the center of the Earth is not an ideal Fermi gas of electrons. However, in the realm of astrophysics, such gases do occur inside white dwarf stars. Fermi then went on to give estimates of the melting point of iron at such pressures on the assumption that a solid melts when the thermal vibrations of its atoms reach a certain fraction of its interatomic distance.

Such questions are not discussed in Fermi's lecture notes. We have included this aside to show how involved Fermi was in geophysics questions at the time of the lectures.

14.3 Estimates of the Earth's Interior Density

There is a long history of attempting to describe the Earth's potential energy by models of the planet's shape and density distribution. The items of data that were used, most of them discussed earlier in these notes, are: (1) total mass; (2) ellipticity of the meridian; (3) difference of moments of inertia; and (4) density at surface. Many analytic parameterizations of the Earth's density were given over the years. The reader may recall that in Sec. (2.3), Fermi wrote down a simple formula for $\rho(r)$ in Eq. (2.38)). Two nineteenth-century examples, the first by Edouard Roche and the second by Friedrich Helmhert (Brush 1996) are presented below, with ρ measured in g/cm^3 and R_e the Earth's radius. Roche's formula is

$$\rho = \rho_0[1 - \beta r^2/R_e^2], \tag{14.2}$$

where $\rho_0 = 10.1, \beta = 0.764$, and Helmhert's formula is

$$\rho = \rho_0[1 - \beta_1 r^2/R_e^2 + \beta_2 r^4/R_e^4], \tag{14.3}$$

where $\rho_0 = 11, \beta_1 = 1.04, \beta_2 = 0.275$.

Such parameterizations assume that the Earth's density varies smoothly, as if there were no discontinuities in its chemical composition. Such is not the case as is seen by the abrupt change in the impedances[1] of seismic waves that takes place

[1] The impedance I of a seismic wave of velocity v in a medium of density ρ, is $I = \rho v$.

at a depth of approximately 2900 m below sea level. Known as the Gutenberg discontinuity, so named after the late German geophysicist Beno Gutenberg, this corresponds to the boundary between the Earth's solid mantle and its liquid core. Fermi also quotes data showing the rise in density in g/cm^3 to be linear from a value of 3.4 at a depth of 60 km to one of 4.7 at 1200 km and again linear from 4.7 at 2450 km to 5.0 at 2900 km, the point of the Gutenberg discontinuity (Brush 1996). From there until the Earth's center, which lies at a depth of 6371 km, he quotes a constant value of $11.0\,g/cm^3$. This, as noted earlier, is less than the modern value of the density of the Earth in the inner core, $12.0 - 13.0\,g/cm^3$.

14.4 Early Estimates of the Age of the Earth

The discovery of radioactivity near the turn of the twentieth century had a profound and immediate effect on geophysics, most notably on the ongoing debate of the Earth's age. Early estimates of this number, principally those of Lord Kelvin, followed from calculations of heat conduction and were based on observed temperature gradients near the Earth's surface. Depending on the assumptions made, the number obtained varied from twenty million years to upwards of a hundred million years, insufficient to account for the observed time of geological changes and certainly too short for the evolution of species by natural selection proposed by Charles Darwin and Alfred Wallace. Radioactivity showed that highly significant and previously unaccounted sources of heat were likely to be present in the Earth's crust and that, as has proved the case, any calculation needed to take into account these new sources (Jackson 2006).

14.5 Radioactive Decay Equation

Within a relatively short time, the work of Lord Rutherford and others determined that radioactive decay was due to the spontaneous disintegration of atomic nuclei and could be characterized by the emission of alpha, beta, and gamma rays (Cook 1973). These were respectively due to a nucleus emitting a helium nucleus, an electron, or a photon. It was quickly seen that in all cases the rate of decay was simply proportional to N, the number of nuclei present in the original sample,

$$dN/dt = -\lambda N, \tag{14.4}$$

where λ was known as the decay constant. Accordingly the number of daughter nuclei grows as

$$dN_1/dt = +\lambda N. \tag{14.5}$$

The solution to Eq. (14.4)is

$$N(t) = N(0)\exp(-\lambda t). \tag{14.6}$$

The half-life of the decaying nucleus, $T_{1/2}$, is the time in which $N(t)$ is reduced to half its original value. It is simply given by

$$\exp(-\lambda T_{1/2}) = 1/2 \tag{14.7}$$

so that $\lambda T_{1/2} = \ln 2$.

14.6 Heat by Radioactivity within the Earth

It is believed that radioactivity is responsible for approximately half of the 47 Tw (1 Tw = 10^{12} watts) flowing to the surface from the Earth's interior, the other half being primordial heat generated by a variety of causes during Earth's formation. While this amount of heat is only 0.027 % of the 173,000 Tw from insolation, the latter only penetrates a few tens of centimeters into the Earth. Primordial heat and radioactivity are responsible for observed geological processes such as plate tectonics and volcanic activity.

The key radioactive elements are thorium; the two isotopes of uranium; and the rare isotope of potassium, ^{40}K. This last isotope constitutes some 0.012% of the potassium found in nature and decays with a long life-time: 1.25×10^9 years. Its two principal decay modes are to ^{40}Ca, accounting for 89% of the decays, and ^{40}Ar, for the remaining 11%. Though comparatively rare, this potassium isotope is important in the Earth's heat production because potassium is relatively abundant, of the order of a few percent in both granite and basalt, the principal components of the Earth's crust. As a side note, this decay is also responsible for the almost 1% of the Earth's atmosphere being composed of argon. The heat generated per second by a gram of each of these radioactive elements is listed in Table (14.1) (this data is from Fermi's notes) and Table (14.2) lists the percentages of these elements in common rocks.

Table 14.1: Rate of heat generation by K, U, Th

K	U	Th
3.9×10^{-12}cal/(g s)	2.5×10^{-8}cal/(g s)	5.6×10^{-9}cal/(g s)

With these numbers Fermi was in a position to estimate the amount of heat generated in the Earth by radioactivity. Multiplying the proportional amount of each radioactive element in a type of rock by the heat generated per gram by that element, we find for the case of granite

Table 14.2: The percentages of radioactive elements in common rocks

rock	K	U	Th
granite	2.6×10^{-2}	6×10^{-6}	2.0×10^{-5}
basalt	1.28×10^{-2}	1.6×10^{-6}	0.6×10^{-5}
peridotite	0.22×10^{-2}	0.3×10^{-6}	0.1×10^{-5}
free iron	0.14×10^{-2}	0.02×10^{-6}	0.06×10^{-5}

$$\begin{aligned} Q_{granite} &= (3.9 \times 10^{-12}) \times (2.6 \times 10^{-2})\ (\text{K}) \\ &+ (2.5 \times 10^{-8}) \times (6 \times 10^{-6})\ (\text{U}) \\ &+ (5.6 \times 10^{-9}) \times (2 \times 10^{-5})\ (\text{Th}) \\ &= 3.6 \times 10^{-13} \text{cal}/(\text{g}\,\text{s}). \end{aligned} \tag{14.8}$$

The biggest unknown lies in estimating how many grams of the material containing the radioactive isotopes are present in the Earth. The value for granite of 3.2×10^{20} cal/year given below in Table (14.3) corresponds to estimating that granite makes up $\approx 0.5\%$ of the Earth's mass, or the total amount of granite is $\sim 3 \times 10^{25}$ g. This is a reasonable estimate given that although granite is the main component of the first few tens of kilometers of the continental crust, there is little of it elsewhere in the Earth. Making similar estimates for the other types of rock and multiplying our results by the number of seconds in a year, 3.15×10^7, Fermi obtained estimates of the amount of heat generated in the Earth measured in units of cal/year for each of the four types of rock. These are listed in Table (14.3). Given the relative scarcity of radioactive elements in peridotite, it seems surprising at first glance that it is the principal contributor to the heat balance. The answer is quite simple: peridotite is the dominant rock in the Earth's mantle and therefore constitutes a far larger compo-

Table 14.3: Heat production in common rocks

granite	basalt	peridotite	free iron
3.2×10^{20} cal/year	2.6×10^{20} cal/year	16.1×10^{20} cal/year.	4.3×10^{20} cal/year

nent of the Earth's volume than either granite or basalt, key elements of the considerably smaller crust of the Earth.

To appreciate the scale of the heating due to radioactivity, imagine that the heat generated by radioactivity remained entirely within the Earth rather than escaping into the atmosphere. The sum of the heat production for the rocks listed in Table (14.3) is $H_{rad} = 26.2 \times 10^{20}$ cal/year. Taking the specific heat of rock as approximately $c_e = 0.2\,\mathrm{cal/(g\,K)}$ we would then expect the Earth's temperature to rise by

$$\frac{H_{rad}}{M_e c_e} = \frac{26.2 \times 10^{20}}{6 \times 10^{27} \times 0.2} = 2.2 \times 10^{-6}\mathrm{K/year}$$

or more than 2000 K in a billion years. This has not happened because heat escapes into the atmosphere.

The methods and the calculations above are correct, but the results reflect data available in 1941. Modern measurements have improved and changed our present understanding of these numbers. In particular the estimate of heat generated by radiation that reaches the Earth's surface, $H_{rad} = 26.2 \times 10^{20}$ cal/year, is almost an order of magnitude larger than the value we presently believe to be true. There are many reasons for the discrepancy, including overly large estimates of the amount of radioactive material present, particularly true in the case of peridotite. Some of the heat also surely is responsible for the movement of the tectonic plates, a phenomenon unknown or at least unaccepted in 1941. In addition, it was believed at the time that the thickness of the crust underneath the oceans and the continents was comparable. We now know that this is not the case. Whereas the continental crust, a material comparatively rich in uranium and thorium, extends for 30 km or so, the oceanic crust has a thickness of only a few kilometers.

14.7 Uranium Decays and Lead Isotope Abundances

In what follows Fermi focuses on three of the four isotopes that contribute significantly to radioactively generated heat in the Earth. Each one of these three, ^{238}U, ^{235}U, and ^{232}Th, has a long decay chain that terminates in one of the four stable isotopes of lead, the lead isotopes having respective atomic weights of 208, 207, 206, and 204. They have significant abundances in common lead (52.4%, 24.1%, 22.1%, and 1.4%, respectively), and all four have primordial components, but only the first three are produced as the end products of radioactive decay by elements of higher atomic number. We list below the three principal modes in question and their respective half-lives.

Table 14.4: Decay Modes of Uranium Isotopes

$^{238}\mathrm{U} \to 8\mathrm{He}^4 + 6\mathrm{e}^- + {}^{206}\mathrm{Pb}$	$T_{1/2} = 4.56 \times 10^9$ year
$^{235}\mathrm{U} \to 7\mathrm{He}^4 + 4\mathrm{e}^- + {}^{207}\mathrm{Pb}$	$T_{1/2} = 7.13 \times 10^8$ year
$^{232}\mathrm{Th} \to 6\mathrm{He}^4 + 4\mathrm{e}^- + {}^{208}\mathrm{Pb}$	$T_{1/2} = 1.31 \times 10^{10}$ year

As indicated, the transition from uranium or thorium to lead comes as the conclusion of many steps that involve either the emission of an electron or of an alpha particle.The half-life measurements given in Fermi's notes differ slightly from the presently determined best values but are close enough for the calculations below.

There is a subtlety here which is worth mentioning. In

principle the decay of, e.g., ^{238}U involves fourteen steps in succession, eight of them involving the emission of an alpha particle and six of them that of an electron. This raises the question of what does λ mean? The ith daughter in the decay chain satisfies

$$dN_{i+1}/dt = \lambda_i N_i - \lambda_{i+1} N_{i+1} \quad (14.9)$$

rather than the the simple form of Eq. (14.5), so it would appear that we have to solve a large set of simultaneous differential equations. Fortunately this is not the case because, in all three of isotope decays listed above, the first decay has a much longer half-life than all the others, or equivalently,

$$\lambda_1 \ll \lambda_2, \lambda_3, \ldots \lambda_n, \quad (14.10)$$

so the $T_{1/2}$ quoted in the table above is the $T_{1/2}$ for the first decay in the chain. This reasoning does not hold for ^{40}K, which has two decay modes with comparable lifetimes, but we will not pursue that case here.

14.8 Using Radioactive Isotopes for Dating

There are many examples of employing radioactive isotopes to date rocks or other materials. We shall give one example here. It makes use of zircon, a mineral that during its formation incorporates uranium and thorium but not lead into its crystal structure. Any lead present in an undamaged sample of zircon is therefore the result of uranium or thorium decay.

Let $N_{238}(t)$ be the number of ^{238}U atoms present at time t, where t is the number of years since the rock was formed, i.e., it is the age of the zircon rock. Likewise, let $N_{206}(t)$ be the number of ^{206}Pb atoms present at time t. We know that $N_{238}(0)$, the number of ^{238}U atoms at the time of the zircon rock formation, is equal to $N_{238}(t) + N_{206}(t)$ because originally there was no lead in the zircon and the ^{238}U atoms that decayed are now present in the zircon as ^{206}Pb atoms. Writing this as an equation, we have

$$N_{238}(0) = N_{238}(t) + N_{206}(t), \qquad (14.11)$$

or, dividing by $N_{238}(t)$,

$$\frac{N_{238}(0)}{N_{238}(t)} = 1 + \frac{N_{206}(t)}{N_{238}(t)}. \qquad (14.12)$$

Using the radioactive decay law, Eq. (14.6), we have

$$N_{238}(t) = N_{238}(0)e^{-\lambda_{238}t}, \qquad (14.13)$$

so Eq. (14.12) becomes

$$e^{\lambda_{238}t} = 1 + \frac{N_{206}(t)}{N_{238}(t)}. \qquad (14.14)$$

Taking logarithms of both sides, we obtain

$$t = \left(\frac{\ln(1 + \frac{N_{206}(t)}{N_{238}(t)})}{\lambda_{238}}\right) = \left(\frac{\ln(1 + \frac{N_{206}(t)}{N_{238}(t)})}{\ln(2)}\right)(T_{1/2})_{238}, \qquad (14.15)$$

where from Table (14.4), $(T_{1/2})_{238} = 4.56 \times 10^9$year. The last form of Eq. (14.15) is quite useful. It gives the age t of a sample of zircon as a fraction of the half-life of ^{238}U. The fraction depends only on the ratio of ^{206}Pb to ^{238}U at time t in the sample. Fermi gives an example of such dating to determine the age of geological eras. In Table (14.5), we use his method to calculate the ages of various eras in the Earth's history, using Eq. (14.15) to calculate the age of the era from the Pb/U ratios.

As a check on the results obtained using the decay of ^{238}U to ^{206}Pb, we can instead use the decay of ^{235}U to ^{207}Pb. The

Table 14.5: Eras, Pb/U ratios, and era lengths

Era	Pb/U ratio	Era length (years)
Oldest	0.228	1.4×10^9
Lower Cambrian	0.08	5.1×10^8
Carboniferous	0.04	2.6×10^8
Lower Cenozoic	0.008	5.2×10^7

age of the same sample of zircon would now be given in terms of the ratio of ^{207}Pb to ^{235}U,

$$t = \left(\frac{\ln(1 + \frac{N_{207}(t)}{N_{235}(t)})}{\lambda_{235}} \right) = (T_{1/2})_{235} \left(\frac{\ln(1 + \frac{N_{207}(t)}{N_{235}(t)})}{\ln(2)} \right), \tag{14.16}$$

where from Table (14.4), $(T_{1/2})_{235} = 7.13 \times 10^8$years.

The general method described above is known as U-Pb dating and is one of the oldest and most reliable methods of radiometric dating. As the preceding discussion hopefully makes clear, the discovery of radioactivity and the use of radiometric dating revolutionized the understanding of the age of the Earth. Present estimates are consistent with the time required for biological and geological evolution.

(18)

$\varkappa = 0.005 \qquad \gamma_0 = 3.2\times10^{-4}$ deg/cm

$c = 0.2 \qquad \rho = 3$

Loss of heat (cal/sec)

$$4\pi\,(6.37\times10^{8})^{2}\,0.005\times3.2\times10^{-4} = 8.2\times10^{12}\ \text{cal/sec}$$
$$= 2.6\times10^{20}\ \text{cal/year}$$

Mass of Earth 6×10^{27} heat capacity 1.2×10^{27}

represents a cooling by 1°C in 5×10^{6} years

or 400°C in 2×10^{9} years

Estimate of gradient

$$\varkappa\gamma\,dt = d\left(c\rho\,\frac{T_0^2}{2\gamma}\right)$$

$$t = \frac{c\rho}{4\varkappa}\left(\frac{T_0}{\gamma}\right)^2$$

$$\frac{T_0}{\gamma} = \frac{2000}{3.2\times10^{-4}} = 6.3\times10^{6} \quad (63\ \text{km})$$

$$t = \frac{0.6}{4\times0.005}\left(6.3\times10^{6}\right)^2 = 1.2\times10^{15}\ \text{sec}$$
$$= 40\times10^{6}\ \text{years}$$

Equilibrium with production of heat

$$4\pi r^2\,\varkappa(r)\,\frac{dT}{dr} + 4\pi\int_0^r r^2\,dr\,q(r) = 0$$

Example $\quad q(r) = 0 \quad$ for $r < a-\varepsilon$

$q(r) = q = \text{const}$ for $a > r > a-\varepsilon$

$$\varkappa\frac{dT}{dz} = (\varepsilon - z)\,q$$

Heat flow through Earth's surface

CHAPTER 15

The Physics of Heat Flow in the Earth

This section of Fermi's notes is concerned with the flow of heat to the surface of the Earth from its interior. The sources of this heat can be broken into two broad classes even though it is not possible to completely disentangle them. The two are energy released in radioactive decay and so-called primordial heat. We believe the heat arriving at the Earth's surface from radioactive decay is for the most part generated in the mantle and the Earth's crust. These regions contain isotopes of uranium, thorium, and potassium, the main sources of Earth's radioactivity, while the presence of these isotopes in regions below the mantle is unlikely according to present models of the Earth's interior.

The major components of the primordial heat come from accretion during the Earth's formation and energy released in the establishment of the Earth's iron core. Hence their origins lie in regions that are deeper in the Earth's interior. It is also believed that the flow of heat in the Earth's crust is primarily by conduction, while convection plays an important role in the mantle and the core. Regardless of its source, the mean power delivered to the Earth's surface from sources inside the Earth has been determined to be

$$P_E = 47(\pm 2) \times 10^{12}\,\mathrm{W}. \tag{15.1}$$

The oceanic heat current is typically about 50% higher than the current in the continental crust, the higher value being due to factors such as the magma at floor ridges and the subsequent spreading of the ocean floor.

We can understand the figure for P_E by taking reasonable average values of the continental and oceanic heat currents, namely,

$$q_{cont} = 6.7 \times 10^{-2}\,\mathrm{W/m^2}, \quad q_{ocean} = 10^{-1}\,\mathrm{W/m^2}, \tag{15.2}$$

and multiplying them, respectively, by the continental area of the Earth, $\sim 2 \times 10^8\,\mathrm{km^2}$, and $\sim 3.1 \times 10^8\,\mathrm{km^2}$ for the oceans. This gives

$$\begin{aligned} P_E &= (2 \times 10^{14}\,\mathrm{m^2})(67 \times 10^{-3}\,\mathrm{W/m^2}) \\ &+ (3.1 \times 10^{14}\,\mathrm{m^2})(101 \times 10^{-3}\,\mathrm{W/m^2}) \\ &= 45 \times 10^{12}\,\mathrm{W}. \end{aligned} \tag{15.3}$$

P_E is thousands of times smaller than the power arriving from the Sun. However, solar power affects only the atmosphere and the first several centimeters below the Earth's surface, while the inner Earth heat currents, driven relentlessly forward by the temperature gradient, provide the energy for volcanic activity and all the observed major geologic processes such as the movements of tectonic plates.

15.1 Periodic Heat Variations Near the Earth's Surface

Much of the material in this section was already discussed in Sec. (8.3), but following Fermi, we return to the subject one more time. The same daily and annual changes in the atmosphere's temperature that are responsible for variations below sea level also cause a periodic temperature dependence below ground level. For convenience, we rewrite the solution to the heat conduction equation, Eq. (8.20), as

$$T(t,z) = \bar{T}_0 + \frac{\bar{T}_d - \bar{T}_0}{d} z + \delta T \cos(\omega t - az + \eta) e^{-az}, \quad (15.4)$$

where, as before,

$$a = \sqrt{\frac{\omega_d}{2\chi}}. \quad (15.5)$$

In applying Eq. (15.4) to the temperature variation near the surface of the Earth's crust, the values of the mean surface temperature $\bar{T}_0$ and its periodic variation δT can be set to the same values used over the ocean, but the temperature gradient in the crust $(\bar{T}_d - \bar{T}_0)/d$ is much larger than ($\sim$ 30 K/km), and of the opposite sign to, that of the ocean; in the crust, the temperature increases with increasing depth. However, our main interest is in the value of a, which now depends on the thermal diffusivity of the crust. A reasonable value for it is

$$\chi_{crust} = 8.0 \times 10^{-7}\,\mathrm{m}^2/\mathrm{s}, \quad (15.6)$$

significantly larger than the value of $1.4 \times 10^{-7}\,\mathrm{m}^2/\mathrm{s}$ for water. The resulting values for a and $1/a$ for daily temperature variation in the Earth's crust are

$$a = 6.7\mathrm{m}^{-1}, \text{ or } \frac{1}{a} = 15\,\mathrm{cm}. \quad (15.7)$$

For annual variation, the value of $1/a$ is increased by $\sqrt{365}$, resulting in a value for annual variation of 283 cm.

In the nineteenth century, Lord Kelvin proposed using the periodic variation of temperature at various depths as a way to find $1/a$, and thereby deduce the thermal diffusivity of rock in the Earth's crust and finally obtain its thermal conductivity. However, this turned out to be difficult due to the presence of water in most rocks near the surface.

15.2 Estimate of Heat Passage through the Earth's Surface

The mean continental heat current q is related to the thermal gradient by

$$q_{cont} = \kappa_{crust} \frac{dT}{dz}. \tag{15.8}$$

Fermi used

$$\frac{dT}{dz} \approx 3.2 \times (10^{-4})\,\mathrm{K/cm}, \text{ and } \kappa_{crust} = 0.005\,\mathrm{cal/K\,cm}. \tag{15.9}$$

Converting units, and computing the continental heat current with these parameters, gives

$$q_{cont} = 0.067\,\mathrm{W/m^2},$$

the value we used earlier in our estimate of P_E. Fermi also estimated the value of the total power by multiplying this q current by the area of the Earth, obtaining of course a value that was too low because it was not yet well known that the oceanic heat current is significantly higher than the continental one.

15.3 Space-Time Scales in Heat Conduction

Before a heat-conducting system has reached a steady state or if it is perturbed from a steady state, the temperature of the system obeys the time-dependent heat equation. Since the flow of heat in the Earth's crust is essentially one dimensional, we have

$$\frac{\partial T}{\partial t} = \chi \frac{\partial^2 T}{\partial z^2}, \tag{15.10}$$

Daily experience gives qualitative information about the space and time scales of heat conduction. As an example, suppose a steel rod, initially at room temperature, has its end placed in boiling water. If the rod is touched at an elevation above the boiling water, it will begin to feel "hot" at a time that increases as the elevation of the touched point increases. Using Eq. (15.10), a more quantitative statement can be made. In general the solution of the equation is a function of both z and t, but if we impose boundary conditions that do not set scales,

$$T(z,0) = T_1 \qquad T(0,t) = T_2,$$

with T_1 and T_2 equal to constants, the solution for $T(z,t)$ is a function of only the dimensionless variable

$$V = z^2/(\chi t).$$

Temperature's partial derivatives can then be written as

$$\frac{\partial T}{\partial t} = -\frac{dT}{dV} \cdot \frac{V}{t} \qquad \frac{\partial T}{\partial z} = \frac{dT}{dV} \cdot \frac{2V}{z}.$$

With a little work we see that the heat conduction equation then reduces to a first-order differential equation,

$$\frac{dT}{dV} = \frac{A}{V^{1/2}} \exp(-V/4),$$

where A is a constant. Given this dependence, a semiquantitative criterion for deciding when a region is "hot" is

$$\exp(-V/4) = \exp(-z^2/4\chi t) = \frac{1}{e},$$

or

$$t \sim \frac{z^2}{4\chi}. \tag{15.11}$$

In general, there will be an additional constant multiplying $z^2/(4\chi)$, but ratios of times for different z values will still be captured by Eq. (15.11).

Ordinary steel has $\chi \approx 2 \times 10^{-5}\,\mathrm{m^2/s}$. Table (15.1) gives the time scale from Eq. (15.11) for distances on the rod of 1, 10, and 100 cm. The results illustrate the important point that the time scale goes as the *square* of the corresponding distance scale.

Table 15.1: Heat conduction scales for a steel rod

z (cm)	t (s)
1.0	1.25
10.0	125
100	1.25×10^4

Fermi uses a formula equivalent to Eq. (15.11) to estimate time and distance scales for heat conduction in the Earth. Assume a disturbance in a steady state of heat flow arises at the junction between the Earth's crust and mantle. Setting $z = 0$ at this boundary, and measuring z upward, we may apply Eq. (15.11) to relate distance and time scales for the propagation of this disturbance upward into the crust. For the Earth's crust, a reasonable value is $\chi_{crust} = 8.0 \times 10^{-7}\,\mathrm{m^2/s}$. Using this for χ in Eq. (15.11), several time and distance scales are given in Table (15.2). Table (15.2) again illustrates the very rapid growth with distance of the time for a disturbance to

Table 15.2: Heat conduction scales for the Earth's crust

z	t
1.0 m	3.6 days
100 m	99 years
1 km	9900 years
10 km	990000 years

propagate. Given that the Earth's crust is approximately 35 km thick, a perturbation at the mantle-crust boundary would take over 12 million years to reach the Earth's surface. Fermi uses a formula equivalent to Eq. (15.11) to estimate time and distance scales for heat conduction in the Earth. It reads

$$t = C\rho/4\kappa(T_0/\gamma)^2 = \frac{1}{4\chi}(T_0/\gamma)^2,$$

but his T_0 is the estimated temperature at the crust's base and γ is an average temperature gradient in the crust's rock so that (T_0/γ) is a measure of distance. Fermi has a table which makes the same point as Table (15.2).

15.4 Equilibrium with Production of Heat

Fermi also considers heat conduction in the Earth's continental crust in the presence of a distribution of heat sources as a way to include the presence of heat generated by radioactive decays. The one-dimensional heat equation in this case is modified to

$$\rho C_p \frac{\partial T}{\partial t} = \kappa \frac{\partial^2 T}{\partial z^2} + \rho H, \qquad (15.12)$$

where H is the strength of the heat source within the continental crust, with units of power/mass; ρ is the mass density.

For a steady state, we have

$$\kappa \frac{\partial^2 T}{\partial z^2} + \rho H = 0. \tag{15.13}$$

It is reasonable to assume that κ and ρ are independent of z in the continental crust. For the present, we will follow Fermi and also assume that H is independent of z. (This point is discussed further below.) With κ, ρ, and H all constant, integrating Eq. (15.13) gives

$$\kappa \frac{dT}{dz}(z) = \kappa \frac{dT}{dz}(0) - \rho H z. \tag{15.14}$$

Eq. (15.14) relates the heat current at depth z to the heat current at the surface. Denoting the thickness of the Earth's continental crust by ϵ, Eq. (15.14) at $z = \epsilon$ gives

$$\kappa \frac{dT}{dz}(\epsilon) = \kappa \frac{dT}{dz}(0) - \rho H \epsilon. \tag{15.15}$$

The left-hand side of this equation is the heat current entering the crust from the mantle. This is certainly nonzero, but again following Fermi, we temporarily assume it is zero, so all the heat which arrives at the Earth's surface is generated by sources in the crust. This leads to

$$\kappa \frac{dT}{dz}(0) = \rho H \epsilon. \tag{15.16}$$

Fermi uses this equation to estimate ϵ, the thickness of the Earth's crust, using the parameters of section (1.2), namely,

$$\kappa \frac{dT}{dz}(0) = 0.067\,\mathrm{W/m^2}. \tag{15.17}$$

To obtain a value for$H$, the heat content of the rock in the crust, we return to the discussion in section (14.6) of heat generated within the crust by radioactivity. Averaging over granite and basalt, Fermi assumes a value for ρH of $5 \times 10^{-13}\,\mathrm{cal/cm^3\,s}$, or $2.1 \times 10^{-6}\,\mathrm{W/m^3}$. Solving for ϵ, we obtain

$$\begin{aligned}\epsilon &= \kappa \frac{dT}{dz}(0)/(\rho H) \\ &= (0.067\,\mathrm{W/m^2})/(2.1 \times 10^{-6}\,\mathrm{W/m^3} = 31.9\,\mathrm{km},\end{aligned} \tag{15.18}$$

a somewhat small, but not altogether unreasonable, value for the continental crust thickness. However the mantle does inject a significant amount of heat into the crust. As a result, the amount of heat generated by sources in the crust cannot add up to the total heat current at the surface. In addition, although the density of radioactive sources in the mantle is less than that in the crust, the volume of the mantle is approximately a thousand times larger than that of the crust so that most of heat from radioactive decays is generated in the mantle. This should also be expected since, as we have already noticed, despite the continental crust being far thicker than the oceanic one, their heat currents are comparable.

The present-day picture of distribution of radioactive heat sources is that approximately 4/5 of the radioactive heat generated in the Earth comes from the mantle, with only 1/5 coming from the crust (Fowler 1990). A consistent picture can therefore only be obtained by dropping the assumption of a uniform density of sources in the crust and instead allowing a distribution of heat sources in the crust to decrease with increasing depth. In particular the formula

$$H(z) = H(0)\exp(-\frac{z}{h_r}) \tag{15.19}$$

explains the data in many parts of the world (Masters and Constable), with $h_r \sim 10\,\mathrm{km}$. This, of course, incorporates some of the vast amount of information that has been obtained in the 80 years since Fermi's lectures.

Finally, let us discuss the temperature in the Earth's continental crust. We return to Eq. (15.13) and integrate twice, allowing H to depend on z. This results in the following equation for the temperature at depth z,

$$T(z) - T(0) = z\frac{dT}{dz}(0) - \frac{1}{\kappa}\int_0^z dz' \int_0^{z'} dz'' \rho H(z'') \quad (15.20)$$

We are interested in the temperature at the bottom of the Earth's continental crust, at $z = \epsilon$. For an assumed value of $\epsilon = 35\,\mathrm{km}$, and with the previously used value of $dT/dz(0) = 32\,\mathrm{K/km}$, the first term on the right-hand side of Eq. (15.20) is 1120 K. This is the value which would hold in the absence of heat sources, assuming constant heat conductivity. The presence of heat sources in the crust will reduce this number, the amount of reduction depending on the z dependence of $H(z)$. Taking first the case considered by Fermi, i.e., $\rho H(z)$ is a constant equal to $2.1 \times 10^{-6}\,\mathrm{W/m^3}$, and $\kappa = 2.1\,\mathrm{W/m\,K}$, the integral is easily evaluated and and we have

$$T(\epsilon) - T(0) = 1120\,\mathrm{K} - 612\,\mathrm{K} = 508\,\mathrm{K}. \quad (15.21)$$

If on the other hand, we use the model of Eq. (15.19), we obtain

$$T(\epsilon) - T(0) = 1120\,\mathrm{K} - 250\,\mathrm{K} = 870\,\mathrm{K}. \quad (15.22)$$

The actual temperature of any point in the Earth's crust which lies deeper than the deepest borehole is not directly measurable, and is therefore dependent on models for predicted values. From the results just obtained it is reasonable to conclude that in regions of the crust where its depth is near the mean value of 35 km, $T(\epsilon) - T(0)$ should lie between 1120 K (no heat sources in the crust) and 508 K (too many heat sources in the crust).

$\mu = 8.6 \times 10^{25}$ $\quad 8.6 \times 10^{25} / 10^{27} = 0.086$

Assume that there is a magnetic potential

$\operatorname{div} H = 0$

$H = -\operatorname{grad} V \qquad \Delta V = 0 \qquad V = V_i + V_e$

$$V = \sum Y_{lm}(\theta, \varphi) \left[\left(\frac{a}{r}\right)^{l+1} c_{lm} + \left(\frac{r}{a}\right)^{l} \gamma_{lm} \right]$$

$$-Z = \frac{\partial V}{\partial r} = \sum Y_{lm}(\theta, \varphi) \left[-(l+1) \frac{c_{lm}}{r} \left(\frac{a}{r}\right)^{l+1} + l \frac{\gamma_{lm}}{r} \left(\frac{r}{a}\right)^{l} \right]$$

$$V_0 = \sum \left(c_{lm} + \gamma_{lm} \right) Y_{lm}(\theta, \varphi)$$

$$-Z = \sum \left(l \gamma_{lm} - (l+1) c_{lm} \right) Y_{lm}(\theta, \varphi)$$

$$V = a \int_0^{\theta} X \, d\theta \qquad W = U_{\lambda = 0} - a \sin\theta \int_0^{\lambda} Y \, d\lambda$$

$$\frac{V_1}{a} = g_1^0 \cos\theta + g_1^1 \sin\theta \cos\lambda + h_1^1 \sin\theta \sin\lambda$$

	g_1^0	g_1^1	h_1^1
1835 (Gauss)	−3235	−311	625
1885 (Schmidt)	−3168	−222	595
1922 (Dyson)	−3095	−226	592

Convergence is poor

Table page 603

Outside field & field without potential $\quad \int H ds = 4\pi i$

Apparent vertical currents ~ 0.2 amp/Km² ~ 20×10^{-13} amp/cm²

Compare with atmospheric current 1000/500 × 10⁶ amp/Km² = 0.2×10^{-5}

factor 10^5 !;

Local anomalies Kursk p. 611

Secular variations figs p 613–614 period 480 years?

Magnetic field expansion in spherical harmonics

CHAPTER 16

Earth Magnetism

16.1 Units

We briefly discuss the units used in electromagnetism before addressing the question of the Earth's magnetic field. Fermi, like most physicists of his time, used esu-cgs = gaussian-cgs units. In them the gauss (G) is the unit of magnetic field and the now obsolete statampere (statA) is the unit of current. However since World War II, most scientific work on electromagnetism uses SI units in which a magnetic field is measured in tesla (T), where 1 T $= 10^4$ G, and electric current is measured in the familiar ampere (A). It is therefore desirable to use SI units to calculate magnetic fields in tesla from currents in amperes. To then obtain the magnetic field in gauss, simply multiply by 10^4. This two step procedure will be used in what follows.

16.2 Magnetic Moments

The concept of magnetic moment, a vector which we will denote as $\boldsymbol{m}$, plays a key role in the treatment of the Earth's magnetic field. For a planar loop of wire of area $\mathcal{A}$, carrying current I, the magnitude of $\boldsymbol{m}$ is given by $|\boldsymbol{m}| = I\mathcal{A}\,(\mathrm{A}\cdot\mathrm{m}^2)$ regardless of the loop's shape. The direction of $\boldsymbol{m}$ is determined by the sense of the current. If the current flows counterclockwise when viewed from above, $\boldsymbol{m}$ points toward the observer; if the current flows clockwise when viewed from above, $\boldsymbol{m}$ points away from the observer. If a magnetic moment $\boldsymbol{m}$ is placed at the origin, with the z axis parallel or antiparallel to $\boldsymbol{m}$ the magnetic field (in T) is

$$\boldsymbol{B} = \pm(\frac{\mu_0}{4\pi})\frac{|\boldsymbol{m}|}{r^3}(2\cos\theta\hat{\boldsymbol{r}} + \sin\theta\hat{\boldsymbol{\theta}}). \tag{16.1}$$

Here $+$ corresponds to $\boldsymbol{m}$ parallel to $\hat{\boldsymbol{z}}$, and $-$ to $\boldsymbol{m}$ antiparallel to $\hat{\boldsymbol{z}}$. In Eq. (16.1), r and θ are the usual spherical coordinates, while $\hat{\boldsymbol{r}}$ and $\hat{\boldsymbol{\theta}}$ are the corresponding unit vectors. The combination $\mu_0/4\pi$ is given by

$$\mu_0/4\pi = 10^{-7}\mathrm{m}^2\!\cdot\!\mathrm{kg/s}^2\!\cdot\!\mathrm{A}^2.$$

Using this information, a practical formula that leads to the field in gauss from the magnetic moment in $\mathrm{A}\!\cdot\!\mathrm{m}^2$ and r in meters, is

$$\boldsymbol{B} = \pm(\frac{1}{1000})\frac{|\boldsymbol{m}|}{r^3}(2\cos\theta\hat{\boldsymbol{r}} + \sin\theta\hat{\boldsymbol{\theta}}). \tag{16.2}$$

As discussed below in more detail, the case of $\boldsymbol{m}$ antiparallel to $\hat{\boldsymbol{z}}$ is convenient in discussing the Earth's magnetic field. The magnetic moment of the Earth then points from the Northern Hemisphere to the Southern Hemisphere, so the conventional "North Pole" is really a south magnetic pole, or a "north-seeking pole." Taking the magnetic z axis to point toward Greenland (remember the magnetic axis does not coincide with the Earth's axis of rotation), the magnetic moment of the Earth points in the negative z direction, corresponding to the $-$ sign in Eqs. (16.1) and (16.2).

16.3 Magnetic Field of the Earth

The dominant feature of the Earth's magnetic field at locations at or near the Earth's surface is a dipole field. This field can be described by a dipole moment located inside the Earth. There is also a generally much weaker nondipole field, which has been mapped out in fine detail by modern satellite measurements. A challenging inverse problem, still under investigation, consists of trying to determine the features of the electric currents inside the Earth which give rise to the total magnetic field (dipole plus nondipole) observed above its surface.

Restricting the discussion to the dipole field, the magnitude of the dipole moment which gives a good description of the field at the Earth's surface is (Fowler 1990)

$$|\boldsymbol{m}_e| = 7.94 \times 10^{22}\,\mathrm{A{\cdot}m^2}. \tag{16.3}$$

The magnetic dipole vector is at a (downward) angle of approximately 11.5° to the axis of rotation of the Earth. The "magnetic equator" goes around the Earth in a similar manner to the geographic equator, but in such a way that the magnetic moment of the Earth is perpendicular to plane of the magnetic equator. Because of the 11.5° tilt of the Earth's magnetic moment, compass needles point in a direction which deviates somewhat from geographic north. The magnetic field on the Earth's surface also has a vertical component, which can be detected using a "dip needle," which is similar

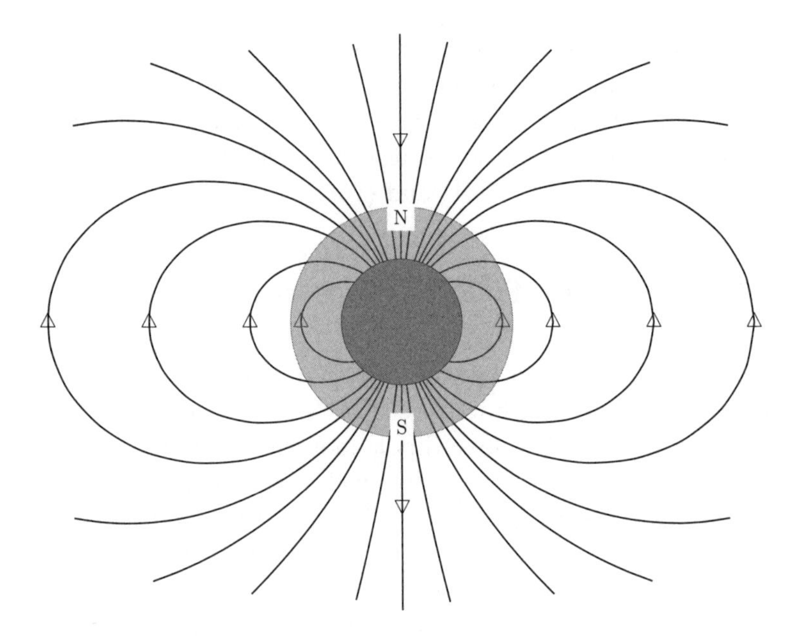

Figure 16.1: Magnetic field lines around Earth

to a compass needle but is allowed to rotate around a horizontal axis instead of a vertical one. At the north pole or south pole the magnetic field has only a vertical component.

The strength of the field at the Earth's surface ranges from 0.25 to 0.65 G with characteristic daily variations at any point of up to $(4-5)\times 10^{-4}$ G. There are also so-called magnetic anomalies, localized differences in the field due to the presence of large iron ore deposits: the region near the Russian city of Kursk is a particularly striking example. But such anomalies do not amount to site-dependent variations of more than a few percent.

Using 6371 km for the radius of the Earth and the value of $|\boldsymbol{m}|$ given above, Eq. (16.2) gives the following formula for the Earth's magnetic field at its surface:

$$\boldsymbol{B} = -(0.31\,\mathrm{G})(2\cos\theta\hat{\boldsymbol{r}} + \sin\theta\hat{\boldsymbol{\theta}}). \tag{16.4}$$

It is valid to think of the $\hat{\boldsymbol{r}}$ and $\hat{\boldsymbol{\theta}}$ components of $\boldsymbol{B}$ as "vertical" and "horizontal." At the magnetic equator ($\theta = \pi/2$), $\boldsymbol{B}$ has only a horizontal component, which points (opposite to $\hat{\boldsymbol{\theta}}$) along a magnetic longitudinal line toward the north pole. At the magnetic north pole, ($\theta = 0$), $\boldsymbol{B}$ has only a vertical component, which points (opposite to $\hat{\boldsymbol{r}}$) downward toward the center of the Earth. Fig. (16.1) shows the pattern of magnetic field lines around the Earth. Note that lines of $\boldsymbol{B}$ emerge from the Earth in the Southern Hemisphere, and enter the Earth in the Northern Hemisphere.

We have not yet addressed what is the cause of the Earth's magnetic field. Largely stimulated by the need for better navigational tools, that field has been a longtime subject of intense study, but its cause has escaped understanding until comparatively recently, chiefly because of ignorance of what lay in the Earth's interior. At the beginning of the twentieth century interest began to focus on mechanical models as a possible source, the principal idea being that the Earth's rotation was the principal agent. Following Fermi, we will briefly discuss two such examples. In both cases, the frequency of the Earth's daily rotation around its axis comes in. This is $\omega = 2\pi/T_{day}$, where the number seconds in a day is $T_{day} = 86,400\,\mathrm{s}$. The numerical value of ω is $7.27\times10^{-5}\,\mathrm{rad/s}$.

16.4 Earth Magnetization and Rotation

The link between angular momentum and magnetization (Galison 1987) was treated by the British physicist Owen W.

Richardson in 1908 and explored experimentally a few years later by a number of authors. In 1915 A. Einstein and W. J. de Haas performed a series of experiments that showed how a freely hanging bar of ferromagnetic material would undergo rotation if the magnetic field in the bar were altered by the shifting of the current in a coil surrounding the bar.

The Barnett effect is the converse of the de Haas Einstein effect. In simple terms, the Barnett effect asserts that if a sample containing angular momentum–bearing atoms or molecules is rotated, there will be magnetic polarization and an effective magnetic field in the frame rotating with the sample. This effect has been observed in laboratory experiments, starting with those of S. J. Barnett himself.

To estimate the magnitude of the effective magnetic field induced by such an effect, begin by considering an electron carrying an angular momentum $\boldsymbol{L}$ that is caused by it traveling in an orbit around a nucleus. Both classically and quantum-mechanically, this motion induces a magnetic moment

$$\boldsymbol{m} = \frac{-e}{2m_e}\boldsymbol{L}, \tag{16.5}$$

where e is the magnitude of the electron's charge and m_e is its mass. As in Eq. (2.64), the Earth rotating with angular velocity ω leads to $\boldsymbol{L}$ in the body-fixed frame satisfying the equation

$$\frac{d\boldsymbol{L}}{dt} = -\boldsymbol{\omega} \wedge \boldsymbol{L}. \tag{16.6}$$

Using the expression above for $\boldsymbol{m}$, this equation becomes

$$\frac{d\boldsymbol{L}}{dt} = \boldsymbol{m} \wedge (\frac{2m_e}{e}\boldsymbol{\omega}). \tag{16.7}$$

From the angular equation of motion, the right side of Eq. (16.7) is the torque acting on $\boldsymbol{L}$. Further, since the torque acting on a magnetic moment $\boldsymbol{m}$ in the presence of an external magnetic field $\boldsymbol{B}$ is $\boldsymbol{m} \wedge \boldsymbol{B}$, we see that the effect of the Earth's rotation is equivalent to a torque exerted on $\boldsymbol{m}$ by a magnetic field

$$\boldsymbol{B} = (\frac{2m_e}{e})\boldsymbol{\omega}. \tag{16.8}$$

This field's magnitude is

$$\begin{aligned} |\boldsymbol{B}| = (\frac{2m_e}{e})|\boldsymbol{\omega}| = 2(\frac{9.1 \times 10^{-31}\text{kg}}{1.602 \times 10^{-19}C})(7.27 \times 10^{-5}\,\text{rad/s}) \\ \sim 8.26 \times 10^{-16}\text{T} = 8.26 \times 10^{-12}\,\text{G}. \end{aligned} \tag{16.9}$$

This is utterly negligible compared to the normal magnetic field of the Earth. Therefore the Barnett effect cannot begin to explain the Earth's magnetic field.

16.5 Rotation of Charges

Fermi also considers the possibility of a magnetic field that is generated by an electric charge density Σ uniformly distributed over the Earth's surface. Such a charge density would lead to an electric field of magnitude $E = 4\pi\Sigma$ directed radially outward or inward, depending on the sign of Σ. From the viewpoint of an observer in a space-fixed frame (one not rotating with the Earth), the charge on the Earth will be in motion, so a space-fixed observer will see a current and therefore a magnetic field. However to an Earth-bound observer, the surface charge on the Earth is not in motion and therefore does not create a magnetic field. Hence the Earth's magnetic field that deflects needles in compasses held by navigators on Earth does not arise from the surface charge of the Earth.

It is of interest to estimate the magnetic field arising from the rotation of the charged Earth as seen by a space-fixed (nonrotating) observer. A simple way to do this is to first compare the magnetic moment due to the rotating charged Earth with the magnetic moment that produces the magnetic field that acts on compass needles for an observer on the Earth. For the latter, recall from Eq. (16.3) that $|\boldsymbol{m}_e| = 7.94 \times 10^{22}\,\mathrm{A{\cdot}m^2}$. Let the magnetic moment due to rotation of the charged Earth be denoted as $\boldsymbol{m}_{rot}$. If the Earth carries a total charge Q_e, and is rotating at angular velocity ω, then an elementary calculation gives

$$|\boldsymbol{m}_{rot}| = \frac{1}{3}|Q_e|\omega R_e^2, \tag{16.10}$$

where R_e is the Earth's radius. As discussed in Sec. (17.2), there is strong evidence that the Earth is negatively charged with $|Q_e| \sim 5 \times 10^5\,\mathrm{C}$. Using $R_e = 6371\,\mathrm{km}$, and $\omega = 7.27 \times 10^{-5}\,\mathrm{rad/s}$, we have

$$|\boldsymbol{m}_{rot}| = \frac{1}{3}|Q_e|\omega R_e^2 \sim 5 \times 10^{14}\,\mathrm{A{\cdot}m^2}. \tag{16.11}$$

From this result we see that $|\boldsymbol{m}_{rot}|$ is at least seven orders of magnitude smaller than $|\boldsymbol{m}_e|$. It would produce a magnetic field which is also seven orders of magnitude smaller compared to the field from $\boldsymbol{m}_e$. So as a matter of principle, to a space-fixed observer, there is a tiny magnetic field arising from the rotation of the charged Earth. However, at a magnitude of roughly one ten-millionth of a gauss, it is unlikely that it will ever be observed.

16.6 Currents inside the Earth

What then causes the Earth's large magnetic moment? It has become clear that the answer lies in the generation of a geodynamo within the Earth's core, a region that extends to approximately 3400 km from the center of the Earth. The inner core, lying in the first 1200 km, is predominantly iron with smaller admixtures of other substances. It is solid because of the enormous pressure there, somewhat more than three million Atmospheres. The next 2200 km, mainly iron and an iron-nickel alloy, are liquid as can be seen, following our discussion in Sec. (12.10), by the fact that transverse seismic waves do not propagate there. The cause of the liquefaction is the high temperature in the core which is on the order of 6000 K, due to the the primordial heat generated during the Earth's formation.

This core temperature is well above the Curie temperature, the point where a metal such as iron loses the ability to form the large domains that can add up to macroscopic magnetic moments. Given that ordinary ferromagnetism is ruled out, the explanation of the Earth's large magnetic moments must lie in the presence of electric currents inside the liquid outer core. These fields would produce magnetic fields via Ampere's law. Since currents are rapidly damped out in a conductor's interior, the necessary currents must flow near the boundary of the conductor, a region known as the conductor's "skin depth" (Jackson 1999). For the Earth, this would imply that the current producing the Earth's magnetic field is concentrated at approximately 3400 km from the Earth's center. To get an idea of the currents required, imagine that the crude model of the Earth's magnetic moment arises from a wire of radius 3400 km carrying a current I. Using the simple formula $|\boldsymbol{m}| = I\mathcal{A}$, we can solve for I, setting

$$I = \frac{|\boldsymbol{m}|}{\mathcal{A}} = \frac{|\boldsymbol{m}|}{\pi R_{core}^2}. \qquad (16.12)$$

Substituting for $|\boldsymbol{m}|$ from Eq. (16.3) and using $R_{core} = 3400\,\text{km}$, we obtain $I = 2.19 \times 10^9\,\text{A}$, a truly enormous current. Admittedly this model is crude but the conclusion that there are enormous currents in the interior of the Earth appears inescapable.

As a point of interest, continue with the single wire model and ask for the sense of the current. Imagine an observer located above the Earth's axis of symmetry in the Northern Hemisphere. To such an observer, the Earth rotates counterclockwise; the Sun rises in the east and sets in the west. Now as discussed earlier, the magnetic moment of the Earth points away from such an observer. The current that produces a moment pointing away from this observer must flow clockwise, or in a westerly direction, opposite to the sense of the Earth's rotation.

The whole picture becomes even more curious when we take into account the movements of the Earth's magnetic field. Though significant, they are not so large and rapid as to invalidate the use of a compass. Among these changes are the secular variations that take place over the course of a year. The most recent example is a decline in intensity of the order of 0.05% per year and a westward drift of the north magnetic pole by about 0.2° per year. But the most dramatic change is the complete reversal of the field, as seen in examinations of the geological record. These reversals have occurred repeatedly at random times, the interval between them ranging from one hundred thousand to fifty million years with an average of approximately 250,000 years; the latest reversal took place 780,000 years ago. The reversals indicate that over a relatively short period of time the currents producing the Earth's magnetic field have gone from flowing in the same sense as the Earth's rotation to flowing in the opposite sense.

16.7 Magnetosphere

The Earth's magnetic field is assumed to first approximation to be that of a dipole, but high above the Earth's surface the field is strongly distorted by the "solar wind," a continuous stream of high-energy particles principally produced in the Sun's corona. The stream consists mainly of electrons and protons though some alpha particles are also present. The wind's effect on Earth would be devastating were it not at least partially turned aside as it approaches Earth by the Earth's magnetic field. This occurs in the field's upper reaches, a zone known as the magnetosphere. Because of the solar wind, the Earth's magnetic field takes a shape described by some as resembling water streaming around a rock.

On the day side of the Earth the magnetic field is flattened out in a region known as the magnetopause, located at a distance on the order of ten Earth radii. The field's pressure is balanced there by that of the solar wind. Conversely, on the night side, the wind acts to draw the field in a tail that extends past the Moon, 50 or more Earth radii out.

16.8 Magnetic Storms

A geomagnetic storm is a sudden disturbance in the Earth's magnetic field caused by a shift in intensity of the solar wind, such as what occurs during a solar flare. Despite the action of the Earth's magnetic field some particles from the solar wind do enter the Earth's atmosphere. Their entry is more likely to take place within a few degrees of the magnetic poles because the direction of the Earth's field is then closer to that of the particles so that the Lorentz force is less likely to turn them away. In the collisions with oxygen or nitrogen molecules at elevations typically of the order of 100 km, a bright light is often produced. This phenomenon is known as aurora borealis (a term coined by Galileo) near the magnetic north pole and aurora australis near the southern magnetic pole (Campbell 2003).

16.9 Magnetic Potential Expansion

The potential V of the Earth's magnetic field, defined by $\boldsymbol{B} = -\boldsymbol{\nabla} V$, satisfies $\nabla^2 V = 0$ because $\boldsymbol{\nabla} \cdot \boldsymbol{B} = 0$. As first proposed by the great German mathematician Carl Friedrich Gauss, it can therefore be expanded in spherical harmonics:

$$V(r, \theta, \phi) = a \sum_{l=1}^{\infty} \sum_{m=-l}^{m=l} Y_{lm}(\theta, \phi)[(a/r)^{l+1} c_{lm} + (r/a)^l d_{lm}], \tag{16.13}$$

where we have assumed the Earth to be a sphere of radius a.

The components of the $\boldsymbol{B}$ field in spherical coordinates are then

$$Z = -\frac{\partial V}{\partial r} \tag{16.14}$$

$$X = -\frac{\partial V}{r \partial \theta} \tag{16.15}$$

$$Y = -\frac{\partial V}{r \sin\theta \partial \phi}. \tag{16.16}$$

Furthermore, since the sources of the Earth's magnetic field essentially lie within the Earth, the potential can be fitted by setting all the d_{lm} equal to zero. In this case the $l = 1$ term is

$$V_1 = 1/r^2 [A \cos\theta + B \sin\theta \cos\phi + C \sin\theta \sin\phi]. \tag{16.17}$$

The first term in the above potential corresponds to a magnetic dipole field directed along the Earth's axis of rotation, originating from a source at the center of the Earth. The terms proportional to $\sin\theta$ modify the inclination of the dipole but leave its location at the Earth's center unchanged. A more accurate fit to the Earth's magnetic field requires higher terms in Eq. (16.13) since the best dipole fits to $\boldsymbol{B}$ suggest a location of the dipole 300 km or so away from the Earth's center. However the convergence of the spherical harmonic expansion is rather poor, suggesting that a better fit might be obtained by considering the magnetic field as due to multiple dipole fields placed at different locations within the Earth. This would correspond to the field being generated by a number of current loops (Jackson 1999).

Atmospheric electricity

Conductivity

10^{10} Seawater

$10^{7}-10^{8}$ Damp ground

$10^{5}-10^{7}$ Dry ground

$2\times10^{-4}\ sec^{-1}$ Air (sea level)

10^{9} air at 80 km

} electrostatic units

Surface 5.1×10^{18}

$E_0 = -120\ V/m = 0.004\ ESU$ 120 000 × 3.8

Conduction to the Earth 1360 ampere

Total charge of earth/cm² $\frac{0.004}{4\pi} = 3.2\times10^{-4}$ ESU = surf density

$-1.63\times10^{15} = -5.4\times10^{5}$ Coulomb total charge

$$\Lambda = \Lambda_0 e^{\frac{h}{3.8\ km}}$$

$$\Lambda E = \text{constant} \qquad E = E_0 e^{-\frac{h}{3.8}}$$

$$\delta V = 3800 \times 120 \text{ volts} = 4.6\times10^{5} \text{ volts}$$

Due to electrode effect instead

$$\delta V \sim 200{,}000 \text{ volts}$$

Table on page 262

Electrons form negative small ions (adding ~10 molecules)

Positive ions add 10 H_2O molecules.

Large ions don't contribute to conduction but to space charge

Characteristic properties of atmospheric electricity

CHAPTER 17

Atmospheric Electricity

17.1 Overview

The topic of atmospheric electricity is a very broad one with varied phenomena taking place at every layer of the Earth's atmosphere, from the surface of the Earth to the exosphere. As elsewhere, our discussion will largely follow the schematic presentation of Fermi's notes. The terms "electrosphere" and "ionosphere" give a very broad classification of two regions with different electromagnetic phenomena. The first refers to the electric charge on the Earth and the compensating opposite-sign charge that resides in the troposphere and the lower part of the stratosphere. The ionosphere is a much higher-altitude region whose defining property is its reflection of radio waves. Fig. (17.1) shows the layers of the atmosphere and their overlap with the ionosphere. The latter is essentially a plasma; an ionized gas. The electrosphere, on the other hand, contains ions, but these ions are only a small fraction of the particles present.

17.2 Properties of the Electrosphere

In 1795 the French physicist Charles-Augustin de Coulomb discovered that ordinary air conducts electricity, where by "ordinary air," we mean the lower regions of the atmosphere. This was followed in the nineteenth century by the realization that the Earth is negatively charged. In order to have electrical neutrality the British polymath scientist Lord Kelvin proposed that a positively charged region exists at some elevation in the atmosphere. Taking account of Coulomb's result that ordinary air conducts electricity, we thus see that the electrosphere resembles a spherical condenser or capacitor, necessarily a "leaky" one in which an electric current flows downward from the positive region in the atmosphere to the negative one at the Earth's surface. This downward current, by now a well established fact, would cause the Earth's charge to disappear in a matter of minutes if it were not being replenished. The British physicist Charles T. R. Wilson proposed that the positive charge in the atmosphere and the negative charge on the Earth are maintained by the existence of thunderstorms reversing the flow of charge due to the "fine weather" current by creating an atmospheric circuit (Chalmers 1967). This is made plausible by the fact that on an average day the Earth is experiencing 1800 thunderstorms at any time or some 40,000 or more over a 24-hour period.

The Charge on the Earth The average value of the electric field near the Earth's surface is approximately $-100\,\mathrm{V/m}$, where the minus sign takes account of the fact that the field points toward the Earth. (Fermi used the somewhat larger value of $-120\,\mathrm{V/m}$.) The magnitude of the field is related to the charge density on the Earth's surface by Gauss's law. Writing this law in the esu units favored by Fermi and many other theorists, we have

$$E(\text{esu}) = 4\pi\Sigma(\text{esu}), \tag{17.1}$$

where Σ is the charge/area on the Earth's surface. To convert the electric field to volts/meter, divide the left side of Eq. (17.1) by 3×10^4; in order to convert the surface charge density to coulombs/square meter multiply the right side of Eq. (17.1) by 3×10^5 (Jackson 1999). The result is

$$\Sigma(\text{C/m}^2) = \frac{1}{4\pi \times 9 \times 10^9} E(\text{V/m}). \tag{17.2}$$

For $E = -100\,\text{V/m}$, we find

$$\Sigma = -8.85 \times 10^{-10}\,\text{C/m}^2, \tag{17.3}$$

or on the order of 10^{10} electron charges per square meter. We obtain for the total charge on the Earth, using $R_e = 6371\,\text{km}$

$$Q_e = 4\pi(R_e)^2\Sigma = -452\,\text{kC}. \tag{17.4}$$

Using other values of the average electric field obviously gives somewhat different numbers for Q_e, but all estimates agree that the charge is in the vicinity of $-500\,\text{kC}$.

Current Flowing to the Earth If the physical picture of a leaky spherical condenser with charge constantly being replenished is correct, the current should be constant at different altitudes. A constant Earthward current has indeed been observed in many different experiments at altitudes ranging up to 28 km (Holzworth 1991). The average current density is $J_z = 2.4 \pm 0.4\,\text{pA/m}^2$. The total current being delivered to the Earth is

$$I_e = 4\pi R_e^2 J_z \approx 1200\,\text{A}. \tag{17.5}$$

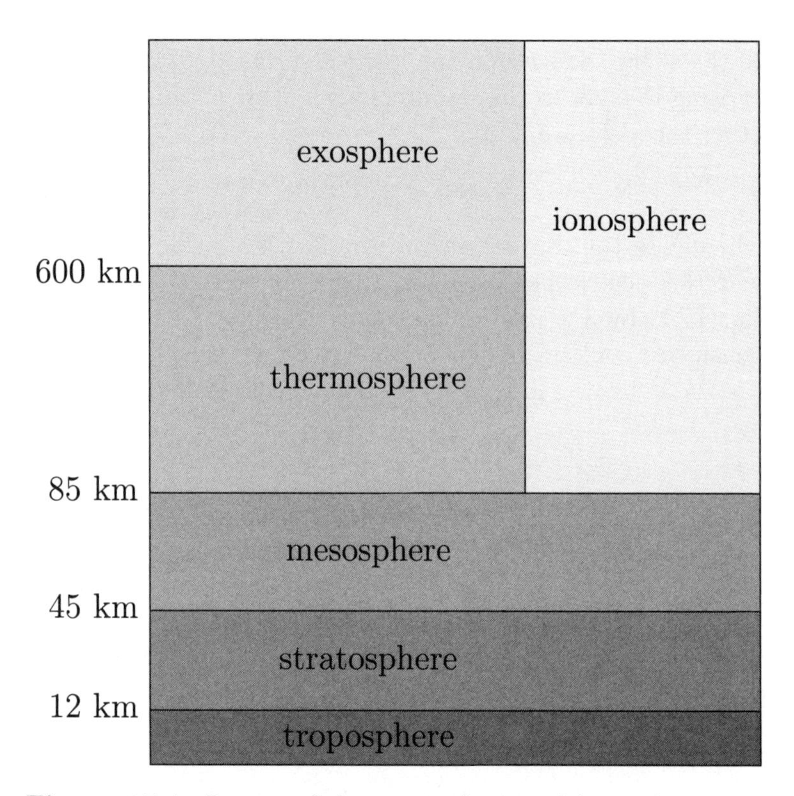

Figure 17.1: Layers of the atmosphere and ionosphere

The time it takes to deliver a charge of Q_e to Earth is

$$T_e = \frac{Q_e}{I_e} \approx 370\,\mathrm{s}, \tag{17.6}$$

or about 6 minutes. Although it does not specifically verify the proposed thunderstorm mechanism, the fact that the downward current density has been observed to remain constant in time in many different experiments is strong evidence that the charge in the atmosphere is constantly being replenished.

Ions in the Electrosphere Experiments in the early twentieth century by C. T. R. Wilson and others established the existence of charged ions in the atmosphere. The various types of ions that exist, most carrying a single unit of electric charge, go by the prosaic names of "small," "intermediate," and "large." We will discuss only the small and large ions.

Either an ion or a neutral particle undergoes many collisions in the atmosphere; these collisions act to change the direction of the particle's velocity thousands of times per second. Charged particles will have an additional drift velocity in addition to this random velocity when an electric field is present. This drift velocity v_d is in the same or opposite direction as the electric field, depending on the sign of the particle's charge. The mobility k is the ratio of the magnitude of the drift velocity to that of the electric field,

$$|v_d| = k|E|, \quad \text{or } k = |\frac{v_d}{E}|. \tag{17.7}$$

It is well established that small ions have different mobilities according to whether their charges are positive or negative. We will follow Fermi in taking the mobility of positively charged ions at ground level to be $k_+ \sim 1.4\,\mathrm{cm}^2/(\mathrm{V}\cdot\mathrm{s})$ and that of negatively charged ions to be $k_- \sim 1.9\,\mathrm{cm}^2/(\mathrm{V}\cdot\mathrm{s})$. Large ions have mobilities roughly 500 times smaller than those of small ions.

Small ions may be thought of as several molecules, many of them water molecules, loosely bound together. If the charge is removed, the molecules are no longer bound together (Chalmers 1967). On the other hand, large ions (much greater in size, e.g., aerosols) remain intact if their charge is removed. They are also more numerous near cities and industrial areas and, because of their small mobility, they play a negligible role in the atmospheric conductivity.

Assuming the numbers of positive and negative small ions

to be roughly equal, and that they are singly charged, the conductivity σ is given by[1]

$$\sigma = en(k_+ + k_-), \tag{17.8}$$

where n is the common number of positive and negative small ions per unit volume. Fermi also writes a formula equivalent to Eq. (17.8).

Eq. (17.8) can be used to determine n at ground level, using the conductivity of $1.33 \times 10^{-16}\,\mathrm{A/(V \cdot cm)}$ (Volland 1984). Using the mobilities given above, Eq. (17.8) gives $n \sim 300\,\mathrm{ions/cm^3}$ at ground level. As might be expected, the experimental values of n near the Earth's surface have a fairly wide variation. Still, $300\,\mathrm{ions/cm^3}$ is a typical value for the density of small ions at the Earth's surface.

Sources of Ions in the Electrosphere The reason there are ions in ordinary air was elucidated by a series of ingenious experiments carried out in the early twentieth century. Ultraviolet photons emitted by the Sun with sufficient energy to ionize air molecules are responsible for the existence of the region known as the ionosphere, also sometimes known as the "Heaviside" layer, so named after the British electricity expert Oliver Heaviside. However, as shown in Fig. (17.1), the ionosphere is at much higher altitude than the ordinary atmosphere, and ultraviolet photons from the Sun, though crucial in this layer, play a negligible role in creating ions in the region of the atmosphere nearer to the Earth.

The cause of ions closer to the Earth, in the so-called ordinary air, was found to arise from two sources: radioactive elements, principally radon, whose origins are in the Earth, and cosmic rays. Radon's dominant radioactive action is due

[1] Fermi uses Λ for conductivity.

to it being the only radioactive element that is gaseous at normal pressure and temperature. The additionally mentioned source of ions was shown to be present by experiments at different altitudes as well as by ones that shielded against radioactivity through the use of lead enclosures. This source was seen to be extraterrestrial and is now known as cosmic rays; cosmic rays are composed of highly energetic heavy particles that enter the atmosphere and undergo numerous ionizing collisions with molecules of the atmosphere. Above an altitude of a few kilometers (Fermi's estimate is 3 km), cosmic rays account for the bulk of ions produced in the atmosphere, on average 10 ion pairs per cubic centimeter per second at sea level (Wahlin 1989), and 4 to 5 times more at an altitude of ~ 20 km.

Conductivity and Potential in the Electrosphere In an ordinary condenser, the potential rises steadily in moving from the negative conductor toward the positive conductor. With some modification, this general idea can be applied to the electrosphere. Starting from the surface of the Earth, the potential increases by 100 V for each meter of altitude. Eventually, the rate of change of the potential decreases, and the potential levels off, analogously to when the positive conductor is approached in an ordinary condenser (Wahlin 1994). Fig. (17.2) shows in a highly schematic way the lines of electric field emanating from positive charges in the atmosphere and ending on negative charges on the Earth's surface. The lines of electric field become more widely spaced as the altitude increases, corresponding to the decrease in the magnitude of the electric field. The figure is idealized in that only positive ions are shown, whereas in reality there is only a slight excess of positive ions over negative ions.

All this can be made more explicit by using the data on

electric conductivity. With increasing altitude, the density of air molecules decreases. This lengthens the mean free path and leads to a rapid increase of conductivity. In the upper troposphere and lower stratosphere, the conductivity increases exponentially (Holzworth 1991),

$$\sigma = \sigma_0 \exp(\frac{z}{H}), \tag{17.9}$$

where $H \sim 7.5\,\text{km}$. In the troposphere and stratosphere, ions undergo frequent collisions with neutral molecules, so Ohm's law[2] holds:

$$J_z = \sigma E_z. \tag{17.10}$$

Given the constancy of J_z with altitude, along with the exponential increase of σ, Eq. (17.10) implies that the electric field decreases exponentially,

$$|E_z| = \frac{1}{\sigma}|J_z| \sim \exp(\frac{-z}{H}). \tag{17.11}$$

For altitudes greater than 4 km, the electric potential, while not literally constant, begins to level off (Wahlin 1994). Although only the electric field is directly measurable, an estimate of the electric potential at altitude z can be made as follows. Set

$$E_z(z) = E_z(0)\exp(-\frac{z}{H}), \tag{17.12}$$

and obtain the potential at z by integrating

$$\begin{aligned} V(z) - V(0) &= -\int_0^z E_z(z')dz' \qquad (17.13) \\ &= -E_z(0)H\left(1 - \exp(-\frac{z}{H})\right). \end{aligned}$$

---

[2]To convert the values Fermi gives for conductivity to SI units, divide by 9×10^9.

Using $E_z(0) = -100\,\mathrm{V/m}$, $H = 7.4\,\mathrm{km}$, and $z = 4\,\mathrm{km}$, the integral in Eq. (17.13) gives

$$V(4\,\mathrm{km}) - V(0) \sim 305\,\mathrm{kV}. \tag{17.14}$$

Fermi uses $H = 3.8\,\mathrm{km}$, $E_z(0) = -120\,\mathrm{V/m}$, and estimates the potential at $z = H$ as $V(H) - V(0) \sim E_z(0)H \sim 456\,\mathrm{kV}$. It is safe to say at altitudes of a few kilometers, the potential relative to the Earth is a few hundred kilovolts.

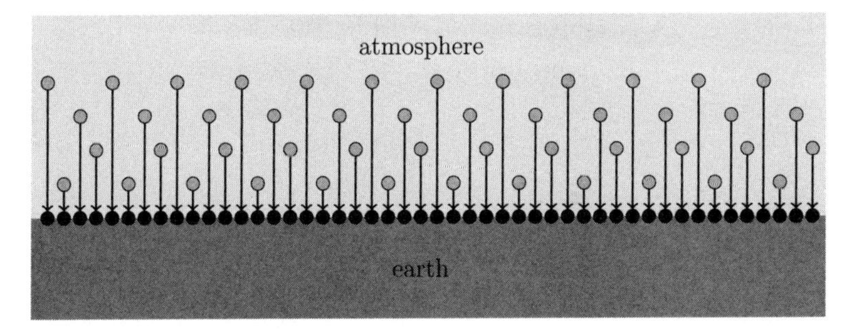

Figure 17.2: Atmospheric charge

Equilibrium The number of ions per unit volume is determined by a balance between sources and sinks. For simplicity, consider an altitude high enough that the electric field is negligible. In such a region, the number of positive ions will be essentially the same as the number of negative ions. Here radioactivity is a small effect, but cosmic rays continually create new ions. Let Π be the number of ions produced per second per unit volume by cosmic rays. Ions can be removed by collisions between positive and negative ions that leave electrically neutral products. If n is the common number of positive and negative ions per unit volume, the rate at which collisions produce neutral products must be proportional to n^2. (This process is called "recombination" in the literature.)

The proportionality coefficient is usually denoted as α. Another way in which a small ion can disappear is to attach itself to a neutral aerosol. Neutral aerosols are sometimes called "nuclei" (definitely not atomic nuclei). We may write the following simple equation for the rate of change of n:

$$\frac{dn}{dt} = \Pi - \alpha n^2 - \beta n N_0, \qquad (17.15)$$

where N_0 is the number of neutral aerosols per unit volume. (Fermi writes a similar equation, but the factor multiplying β is missing.) At equilibrium $dn/dt = 0$, we have

$$\Pi = \alpha n^2 + \beta n N_0. \qquad (17.16)$$

Volland (Volland 1984) gives the following for α and β:

$$\alpha \sim 2 \times 10^{-6}\,\mathrm{cm}^3/\mathrm{s}, \; \beta \sim 1 \times 10^{-5}\,\mathrm{cm}^3/\mathrm{s}. \qquad (17.17)$$

Fermi gives a similar number for α. At an altitude of 15 km, the value of Π due to cosmic rays is $\sim 50\,\mathrm{ions}/(\mathrm{cm}^3 \cdot \mathrm{s})$. If we assume that at this altitude, the number of aerosols is much smaller than n, we can ignore the nN_0 term in Eq. (17.17), and we have

$$\Pi \sim \alpha n^2, \text{ or } 50 \sim 2 \times 10^{-6} n^2, \qquad (17.18)$$

which gives

$$n \sim 5000\,\text{small ions}/\mathrm{cm}^3.$$

This is significantly larger than the sea level value of $n \sim 300\,\text{small ions}/\mathrm{cm}^3$ (Shreve 1970). As cosmic rays enter the atmosphere, more and more are absorbed on the way to the Earth's surface. Due to this absorption, the intensity of cosmic rays increases with increasing altitude. This tracks well with the fact that the density of ions also increases with increasing altitude.

17.3 Water Drops in an Electric Field

In Chap. (5), we discussed the thermodynamic features of liquid drops suspended in a vapor. We examined there the balance between surface tension and pressure difference that stabilizes the drop. That balance becomes more complicated in the presence of an electric field. Since to a first approximation, water may be treated as an electric conductor, there is no electric field inside the drop and therefore no contribution to the energy density of the electric field from the drop's interior. That is not true for the exterior of the drop. There is thus an additional factor that must be taken into account in determining the state of equilibrium between the drop and a surrounding insulating medium.

Under these circumstances does the drop remain spherical or, if not, what shape does it take? This is the problem studied by G. I.Taylor, the great British student of fluid mechanics. He found that an instability, now referred to as the Taylor cone, set in as the electric field intensity grew. The result was that the drop assumes a conical shape whose side has a fixed angle $\theta_0 = 49.3°$ with respect to the axis of symmetry.

To see how this occurs, turn back to Eq. (5.3). We saw there that pure mechanical equilibrium in the absence of an electrical field occurs for a spherical drop when $\delta P = 2\gamma/R$, where γ is the surface tension and R is the drop's radius (we have deliberately used R for radius instead of r in order to avoid later confusion). We also said that there was a more general formula due to Young and Laplace. It consists of replacing $2/R$ with the mean curvature H. In fact the mean curvature becomes singular in the vicinity of a cone's tip, diverging as $1/r$, where r is now taken to be the distance in spherical coordinates from the cone's tip.

The hydrostatic pressure does not diverge as the tip is approached, so equilibrium for a conical surface must instead be obtained by a divergence in the electrostatic field energy that also behaves as $1/r$ if the drop is to take a conical shape. (The presence of a singularity already had been observed in a general way by Benjamin Franklin in his discovery of the lightning rod.) That electrostatic energy is proportional to $\| \boldsymbol{E} \cdot \boldsymbol{E} \|$ or equivalently to $\| \nabla\Phi \cdot \nabla\Phi \|$, where we have written the electric field as the gradient of an electrostatic potential, $\boldsymbol{E} = -\boldsymbol{\nabla}\Phi$.

Since $\boldsymbol{\nabla} \cdot \boldsymbol{E} = 0$, the potential has to satisfy Laplace's equation,

$$\Delta\Phi = 1/r^2(\frac{\partial}{\partial r} r^2 \frac{\partial\Phi}{\partial r} + \frac{1}{\sin\theta}(\frac{\partial}{\partial\theta}(\sin\theta\frac{\partial\Phi}{\partial\theta})) = 0, \quad (17.19)$$

where we have assumed azimuthal symmetry. We look for a solution of the form

$$\Phi(r, \theta) = A(r)B(\theta).$$

The equations to solve are then

$$\frac{\partial}{\partial r}(r^2\frac{\partial A}{\partial r}) = \lambda A$$

and

$$\frac{1}{\sin\theta}(\frac{\partial}{\partial\theta}(\sin\theta\frac{\partial B}{\partial\theta}) = -\lambda B.$$

In order for $\| \boldsymbol{E} \cdot \boldsymbol{E} \|$ to behave as $1/r$ as $r \to 0$, the potential Φ and hence A must behave as $r^{1/2}$ as $r \to 0$.

$A = r^{1/2}$ is in fact an exact solution of the radial differential equation with eigenvalue $\lambda = 3/4$. With that value of λ, the polar angle equation is satisfied by the half-integral Legendre function $P_{1/2}(\cos\theta)$; this function has a single zero at $\theta_0 = 49.3^{\circ}$ with respect to the axis of symmetry. Since the interior of the cone is a conductor, we must have $\Phi = 0$ at the cone's surface. We therefore see that

$$\Phi = Cr^{1/2}P_{1/2}(\cos\theta),$$

with C an overall constant, satisfies all the constraints we have imposed, thus implying that a conducting fluid drop in the presence of an electrostatic field can assume a conical shape at the specified angle. The existence of this singularity has also led to the development of a technique known as electrospraying in which a fine ion emission proceeds from the slightly rounded tip of a fluid cone, but such discussions take us far afield.

(22)

$1.6 \times 10^{-6} \times 600^2$

$1.6 \times 10^{-6} \times 600^2 + 3 \times 10^{-2} \times 600$

Ionisation of high atmosphere

$m\ddot{\xi} = eE$

$-4\pi^2\nu^2 m\xi = eE \qquad \xi = -\frac{eE}{4\pi^2 m\nu^2}$

susceptivity $= \frac{n\xi e}{E} = -\frac{ne^2}{4\pi^2 m\nu^2}$

Index of refraction $\sqrt{\varepsilon} = \sqrt{1+4\pi\chi} = \sqrt{1-\frac{ne^2}{\pi m\nu^2}}$

Reflection at Heaviside-layer

$\frac{1}{\sin\vartheta} = \sqrt{1-\frac{ne^2}{\pi m\nu^2}}$

In vertical direction

$\frac{ne^2}{\pi m\nu^2} = 1 \qquad \lambda = 100\ m \quad \nu = 3\times10^6$

$n = \frac{0.91\times10^{-26}\times9\times10^{12}\pi}{4.8^2\times10^{-20}}$

$= 10^6$

Magnetic effect

$\frac{e}{c}vH = \frac{mv^2}{R} \qquad v = \frac{e}{mc}HR \qquad \frac{v}{2\pi R} = \frac{eH}{2\pi mc}$

$= 2.56\times10^6 H$

Double refraction

Mean specific charge of rain water 0.5 ESU/gram

Lenard-effect $+ \ominus +$

Reflection of electromagnetic waves in the ionosphere

CHAPTER 18

Waves and Plasma in the Earth's High Atmosphere

18.1 Ions in the High Atmosphere

The region Fermi refers to as the high atmosphere, roughly from a little less than a hundred kilometers up to a thousand kilometers above sea level, is also known as the ionosphere, a name given to it because of its electromagnetic properties. The Sun's ultraviolet radiation in this zone is powerful enough to ionize atoms and give the released electrons sufficient energy to prevent them from being immediately recaptured by positive ions. This region's existence was conjectured in 1911 by Oliver Heaviside and A. E. Kennelly in order to explain the apparent bending of transmitted radio waves. Its presence was proved little more than a decade later by E. V. Appleton and M. A. F. Barnett (Bleeker, Geiss, and Huber 2001).

Given the ionosphere's extent, it is not surprising that the details of its behavior are complicated. There are three main layers of increasing altitude: D, $\leq 90\,\mathrm{km}$; E, $90-150\,\mathrm{km}$; and F, $\geq 150\,\mathrm{km}$; each further subdivided. The principal ions in the ionosphere are ionized forms of the atmospheric gases $\mathrm{NO}, \mathrm{O_2}, \mathrm{H}, \mathrm{He}$. The variety of the ions depends on altitude, with $\mathrm{O_2^+}$, $\mathrm{NO^+}$ prevalent in the lower layers while $\mathrm{O^+}, \mathrm{H^+}$ are more abundant in the higher layers. The temperature also varies accordingly, as does the degree of ionization. This depends on the action of the Sun, obviously quite different between night and day, and winter as opposed to summer. The electrons and ions are typically at different temperatures, with electron temperatures of thousands of kelvins in the upper reaches of the ionosphere. Solar flares or flashes of increased brightness also affect the degree of ionization, as does lightning, nor can the effect of cosmic rays be neglected. We shall not go into these details, limiting ourselves instead to a discussion of how it is that the ionospheric plasma leads to the reflection of radio waves.

18.2 Radio Waves and the Electron Plasma

The ionosphere can be treated as an ionized gas or plasma, consisting of electrons, ions, and neutral molecules. We will denote the different charged species by the subscript α, where α takes values e, i for electrons (e) and ions (i). While as mentioned, there are several species of ions, we can capture the relevant physics by assuming a two-fluid plasma in which there are electrons and a single species of ion. The equations of motion for this system consist of Euler-type equations for each of the two fluids, along with Maxwell's equations (Krall and Trivelpiece 1973). As in our discussion of gravity waves in Sec. (9), we will assume that the waves being discussed are small oscillations about an equilibrium state.

In the plasma equilibrium state, there are no external electric or magnetic fields; the electric and magnetic fields that are present are generated by the wave traveling through the plasma itself. As we will see below, the ions are basically immobile. Since the ions and neutral molecules are the main source of mass, gravity plays no role in electromagnetic wave propagation in the ionosphere. Further, the ionosphere is so dilute that its pressure is negligible. As a result, both of the terms on the right-hand side of Eq. (9.3) can be neglected. The force terms in the Euler-type equations for our plasma are strictly electromagnetic in origin. The equations are[1]

$$m_\alpha n_\alpha\left(\frac{\partial \boldsymbol{v}_\alpha}{\partial t} + \boldsymbol{v}_\alpha \cdot \boldsymbol{\nabla} \boldsymbol{v}_\alpha\right) = n_\alpha q_\alpha\left(\boldsymbol{E} + \frac{\boldsymbol{v}_\alpha \times \boldsymbol{B}}{c}\right), \qquad (18.1)$$

where $n_\alpha, m_\alpha, q_\alpha$, and $\boldsymbol{v}_\alpha$ are number density, mass, charge, and velocity for species α. Treating the velocities and electric and magnetic fields as first-order quantities, the term in the magnetic field is the product of two first-order terms and so can be ignored. The equations simplify to

$$m_\alpha n_\alpha \frac{\partial \boldsymbol{v}_\alpha}{\partial t} = n_\alpha q_\alpha \boldsymbol{E}. \qquad (18.2)$$

Assuming sinusoidal time dependence $\exp(-i\omega t)$, we can solve for the $\boldsymbol{v}_\alpha$ in terms of the electric field,

$$\boldsymbol{v}_\alpha = i \frac{q_\alpha \boldsymbol{E}}{m_\alpha \omega}. \qquad (18.3)$$

With the velocities in hand, the total current is

$$\boldsymbol{J} = \sum_\alpha q_\alpha n_\alpha \boldsymbol{v}_\alpha. \qquad (18.4)$$

[1]This section uses esu units.

Substituting in Maxwell's generalization of Ampere's law, we have

$$\nabla \times \boldsymbol{B} = \frac{1}{c}\frac{\partial \boldsymbol{E}}{\partial t} + \frac{4\pi \boldsymbol{J}}{c} = -\frac{i\omega}{c}\boldsymbol{E}\left(1 - \frac{1}{\omega^2}\sum_\alpha \frac{4\pi n_\alpha q_\alpha^2}{m_\alpha}\right). \tag{18.5}$$

At this point, it is useful to simplify the $\sum_\alpha$ term that appears in Eq. (18.5). Assuming only one species of ion, we have $n_e = n_i$, and $q_e = q_i = -e$. Using these notations, we have

$$\sum_\alpha \frac{4\pi n_\alpha q_\alpha^2}{m_\alpha} = \frac{4\pi n_e^2 e^2}{m_e}(1 + \frac{m_e}{m_i}). \tag{18.6}$$

The electron mass is much smaller than any ion mass, so m_e/m_i can be ignored compared to 1. The ions are basically immobile, serving only to provide a neutralizing background to the electron motion. Eq. (18.5) simplifies to

$$\nabla \times \boldsymbol{B} = \frac{1}{c}\frac{\partial \boldsymbol{E}}{\partial t} + \frac{4\pi \boldsymbol{J}}{c} = -\frac{i\omega}{c}\boldsymbol{E}\left(1 - \frac{\omega_p^2}{\omega^2}\right), \tag{18.7}$$

where

$$\omega_p^2 = \frac{4\pi n_e e^2}{m_e}. \tag{18.8}$$

The frequency ω_p is known as the "plasma frequency." To derive a dispersion relation, we will eliminate $\boldsymbol{B}$ using Faraday's law,

$$\nabla \times \boldsymbol{E} = -\frac{1}{c}\frac{\partial \boldsymbol{B}}{\partial t}. \tag{18.9}$$

Taking the curl, we have

$$\nabla \times (\nabla \times \boldsymbol{E}) = -\frac{1}{c}\frac{\partial}{\partial t}(\nabla \times \boldsymbol{B}) = \frac{i\omega}{c}\nabla \times \boldsymbol{B}. \tag{18.10}$$

Using this equation and Eq. (18.7) to eliminate $\boldsymbol{B}$, we have

$$\nabla \times (\nabla \times \boldsymbol{E}) = \frac{\omega^2}{c^2}\boldsymbol{E}(1 - \frac{\omega_p^2}{\omega^2}). \tag{18.11}$$

Taking the spatial dependence in plane wave form $\exp(i\boldsymbol{k}\cdot\boldsymbol{x})$, Eq. (18.11) becomes

$$-\boldsymbol{k}(\boldsymbol{k}\cdot\boldsymbol{E})+k^2\boldsymbol{E}=\frac{\omega^2}{c^2}\boldsymbol{E}(1-\frac{\omega_p^2}{\omega^2}). \tag{18.12}$$

In the study of wave propagation in the ionosphere, an ordinary electromagnetic wave is launched upward from the Earth's atmosphere. Such a wave is transverse, so $\boldsymbol{k}\cdot\boldsymbol{E}=0$. From Eq. (18.12), for $\boldsymbol{E}\neq 0$, a transverse wave must have

$$(kc)^2=\omega^2-\omega_p^2,\ \text{or}\ \omega^2=(kc)^2+\omega_p^2. \tag{18.13}$$

The ionosphere has a number of layers. The electron density and plasma frequency vary with altitude z within the ionosphere, so we have $n_e(z)$ and $\omega_p(z)$. These quantities are measured with a technique called "ionospheric sounding." A radio frequency wave launched from the surface of the Earth will propagate upward into the ionosphere until it reaches an altitude where $k^2=0$. Being unable to travel further in the vertical direction, the wave is then reflected and travels back to the Earth's surface where it is detected. By knowing the frequency and measuring the time delay of the reflected wave, the density of electrons as a function of altitude can be mapped out.

The density of electrons in the ionosphere is found to vary from 10^4 electrons/cm^3 in the lower part of the ionosphere to $\sim 10^{10}$ electrons/cm^3 in the upper reaches of the ionosphere. From Eq. (18.8), the plasma frequency itself varies as the square root of the electron density. At $n_e\sim 10^7$ electrons/cm^3, $\omega_p/2\pi=f_p\sim 30\,\text{MHz}$, a frequency at the boundary between VHF and UHF bands.

18.3 Other Effects

In this section Fermi mentions, without details, a number of other electric effects that occur in the atmosphere.

18.4 Double Refraction

This phenomenon affecting the propagation of electromagnetic waves in the ionosphere is also known as magneto-ionic double refraction. It is is due to the combined effect of the Earth's magnetic field and the plasma of electric charges present in the ionosphere. In our discussion, we will, as before, assume the ions are immobile, and consider only the motion of electrons. In the presence of a static magnetic field, electrons move in circular orbits at the electron cyclotron frequency, ω_c, given by

$$\omega_c = \frac{eB}{m_e c}, \tag{18.14}$$

where B is the external magnetic field on the plasma. In geophysics, the external magnetic field is that of the Earth. Taking the Earth's magnetic field to be $\sim$ 0.5 G, ω_c turns out to be $\sim$ 9 MHz, so in the ionosphere $\omega_c << \omega_p$.

The derivation of the plasma dispersion relation when a magnetic field is present proceeds along similar lines to the derivation of Eq. (18.13). Writing Eq. (18.1) for electrons, again to first order, we now have

$$m_e n_e (\frac{\partial \boldsymbol{v}_e}{\partial t}) = q_e n_e (\boldsymbol{E} + \frac{\boldsymbol{v}_e \times \boldsymbol{B}}{c}). \tag{18.15}$$

In Eq. (18.15), $\boldsymbol{E}$ is, as before, the electric field of the wave itself, a first-order quantity, but $\boldsymbol{B}$ is the external magnetic field, a zeroth-order quantity. Using Eq. (18.15) and Maxwell's equations, dispersion relations can be obtained for the various waves that can propagate. The presence of the magnetic field complicates the derivation since the plasma is no longer isotropic. However, the case where the plasma wave propagates parallel to the magnetic field is relatively simple and will be outlined here.

Even in the presence of the external magnetic field, there is rotational symmetry around the direction of the magnetic field. For this case, there are two dispersion relations, one for right circularly polarized waves and another for left circularly polarized waves. Recall from elementary physics that if the magnetic field points directly toward the observer, the electron moves counterclockwise in a circle at frequency ω_c. For a right circularly polarized wave, the electric field tends to be in phase with the circular motion of the electron. The relation between wave number k_R and frequency for that case turns out to be

$$(k_R c)^2 = \omega^2 - \omega_p^2 \frac{\omega^2}{\omega(\omega - \omega_c)}. \tag{18.16}$$

Note that for frequencies such that $\omega >> \omega_c$ this dispersion relation reduces to that of Eq. (18.13). As ω is reduced, k_R also decreases and eventually reaches zero at a finite frequency. Below this frequency, there is a region where right circularly polarized waves cannot propagate, until the frequency drops below ω_c, where a region of allowed propagation for right circularly polarized waves again opens up.

For left circularly polarized waves, the electric field of the wave tends to be out of phase with the circular motion of the electrons. For this case, the wave number k_L is related to the frequency by the dispersion relation

$$(k_L c)^2 = \omega^2 - \omega_p^2 \frac{\omega^2}{\omega(\omega + \omega_c)}. \tag{18.17}$$

Here again, for frequencies such that $\omega >> \omega_c$ this dispersion relation reduces to that of Eq. (18.13). As ω is reduced, k_L also decreases and eventually reaches zero at a finite frequency. Below this frequency, there is a region where left circularly polarized waves cannot propagate, and for this case this remains true all the way to $\omega = 0$.

Coming down from high frequencies, right circularly polarized waves reach their cutoff frequency before left circularly polarized waves. The difference between the two cutoff frequencies is the cyclotron frequency, ω_c.

Spray Electrification Starting at the end of the twentieth century a series of experiments on the subject were carried out by J. Elster and H. Geitel and soon after that by P. Lenard (Loeb 1958). These led to the conclusion that falling water, whether in the form of raindrops or spray, carried a small electric charge. The phenomenon was also known as the "waterfall effect" since this is where some of the first experiments were carried out. The phenomenon does not have a single simple explanation and, indeed, its causes are numerous and situation-dependent. Since the Earth's surface is generally negatively charged, it seems natural to assume an electrical dipole moment is formed in water drops with the positive charge directed preferentially downward. One can imagine that as drops fall they preferentially repel positively charged ions and attract negatively charged ones. This is but one example of the reasoning one can adopt to begin explaining the phenomenon.

Electrical Storms Electrical storms are commonly known as thunderstorms because the electrical discharges or lightning bolts that characterize them are accompanied by acoustic emissions known as thunder. There is no single explanation for the complicated mechanisms that generate the bolts, nor are these fully understood, but a broad overview says the process starts with the rapid rise of warm, moist air. As it ascends, reaching heights of 15 km or more, it cools and the moisture condenses, forming a cumulonimbus cloud (Feynman, Leighton, and Sands 1966).

Afterword

Enrico Fermi was the only twentieth-century physicist to have reached the very peaks of the profession as both a theorist and an experimentalist. Even setting aside his very considerable experimental work, the breadth of his interests and research as a theorist spanned more fields than any other physicist of his era. One cannot even imagine geniuses such as Einstein, Bohr, Schrodinger, Dirac, Pauli, or Heisenberg teaching a course on geophysics. Fermi doing so is not surprising.

Unfortunately, after the one on which these notes are based, Fermi never taught a course on geophysics again. His work during World War II was entirely devoted to nuclear fission, to the development of a nuclear weapon, and to the problems related to that goal. Though he no longer was a teacher, his ability to grasp the essential features of a problem and his lucidity in explaining possible solutions meant that he continued to be consulted on every facet of those developments. When the war was over, Fermi remained at Los Alamos for a few months during which time he delivered a course on neutron physics. After that, he took a professorship at the University of Chicago.

In the period between 1946 and his early death in November of 1954, Fermi's brilliance in teaching and research made

Chicago a true physics Mecca. Half a dozen of his students went on to win physics Nobel Prizes and most of the others had careers of extraordinary distinction. His research during this time focused largely on nuclear and elementary particle physics as well as astrophysics and cosmic rays. However, he also became a pioneer in computational physics and was one of the first to see the need to study nonlinear phenomena.

Fermi never published any work on geophysics, but he surely kept abreast of the developments in the field. He is known to have made extensive use of what he called his "artificial memory," a set of 3×5 cards that referred to more extensive folders that stored material of particular interest to him. Among them, we find ones on "geophysics" and "geomagnetism and electrical phenomena."

It would not have been surprising to see him study such questions at greater length if his life had not ended at such a tragically young age. Perhaps Fermi might even have once again taught a course on geophysics. In one way, this book allows his teachings on this topic to live on.

APPENDIX A

Solar, Sidereal, and Lunar Days

When the phrase "24 hours a day" is used, the reference is to a solar day. The length of the solar day is determined by two things: the rate at which the Earth rotates around its axis, and the rate at which the Earth moves around the Sun. The sidereal day depends only on the rate at which the Earth rotates around its axis. The two definitions of day can be stated in terms of the positions of astronomical objects relative to the Earth. After one solar day, from the viewpoint of an Earth-based observer, the Sun appears to have returned to the same position in the sky as it was one solar day earlier. After one sidereal day, distant or "fixed" stars appear in the same position they occupied one sidereal day earlier. The sidereal day is approximately 4 minutes shorter than the solar day.

The reason a sidereal day is shorter than a solar day can be seen from Fig. (A.1). The view is looking down on the plane of the Earth's orbit around the Sun (ecliptic plane). For

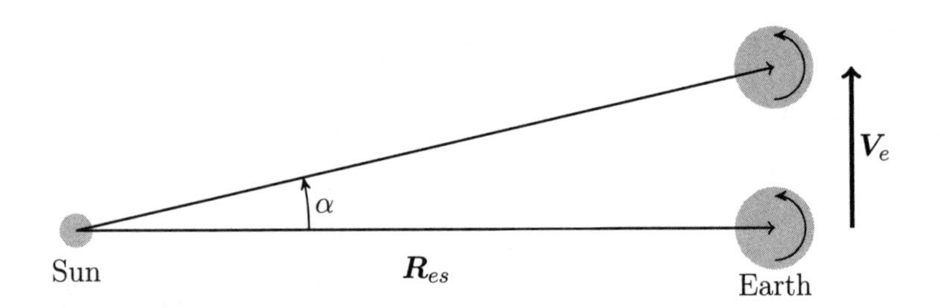

Figure A.1: Solar and sidereal days

simplicity, the tilt of the Earth's axis away from the normal to the ecliptic plane is ignored here, so the Earth's axis is pointing perpendicular to the page. The sense of the Earth's rotation around its axis is counterclockwise, as denoted by the curved arrows. The Earth is moving with a velocity $\boldsymbol{V}_e$ in its orbit around the Sun. It is clear from the figure that for the Sun to appear in the same position as it did one solar day earlier, the Earth must rotate a full 360°, *plus* the angle denoted as α in the figure. The angle α is small, so we can approximate the angle by its tangent and write

$$\alpha \sim \frac{|\boldsymbol{V}_e| T_{day}}{|\boldsymbol{R}_{se}|}. \tag{A.1}$$

Since α is a small angle, it makes little difference whether the sidereal or solar day is used for T_{day} in this formula. Let ω_{se} be the angular frequency of the Earth in its orbit around the Sun. Then we have

$$|\boldsymbol{V}_e| = |\boldsymbol{R}_{se}|\omega_{se} = |\boldsymbol{R}_{se}| \frac{2\pi}{T_{year}}. \tag{A.2}$$

Using this in the formula for α, we have

$$\alpha(\text{rad}) = 2\pi \frac{T_{day}}{T_{year}}, \text{ or } \alpha(\text{deg}) = 360 \frac{T_{day}}{T_{year}} \sim \frac{360}{365} \sim 1 \text{ deg}. \tag{A.3}$$

We conclude that the solar day will be longer than the sidereal day by the time it takes the Earth to rotate 1° around its axis to a good approximation. Now 24 hours corresponds to 360°, so one degree will correspond to (24/360) hour, or 4 minutes. To summarize, the solar day is the time it takes for the Earth to rotate so that the Sun is in the same relative position it was one solar day earlier. This requires the Earth to rotate 361° around its axis. The sidereal day is the time it takes the Earth to rotate around its axis by 360°. The sidereal day is shorter than the solar day by an amount very close to 4 minutes.

Lunar Day

There are two meanings assigned to the term "lunar day" The most common usage refers to the time it takes for the Moon to complete one rotation around its axis. This is approximately 29.5 solar days. The other meaning of lunar day, and the one discussed here, is the time it takes for the Moon to return to the same relative position it was one lunar day earlier. This is the lunar day that affects tides on the Earth.

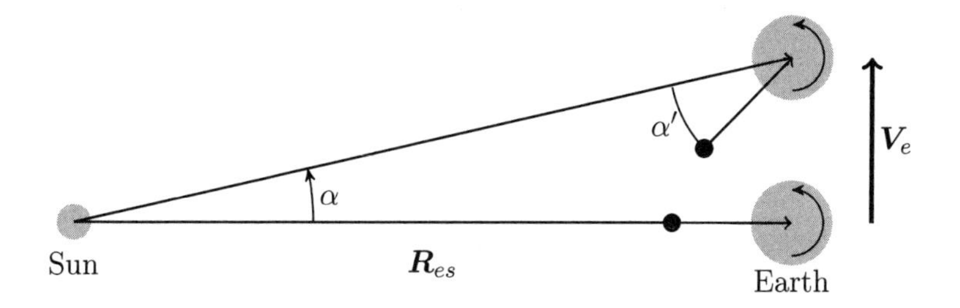

Figure A.2: Lunar day

The length of a lunar day as meant here can be computed in a way very analogous to the discussion of the solar and sidereal days of the previous section. We again ignore the tilt angle of the Earth's axis so the Earth's axis is normal to the ecliptic plane. In addition, we assume the Moon's orbit lies in the ecliptic plane. Fig. (A.2) shows the situation at the time of a solar eclipse, and one solar day later. After one solar day the Moon has moved ahead as shown in the figure by the angle α'. The length of a lunar day is a solar day plus the time it takes the Earth to rotate through the angle α'. At that time, the Moon will be in the same relative position to the Earth that it was one lunar day earlier. To calculate the angle α', we use Eq. (A.3) from the previous section, with T_{year} replaced by T_{month}, where $T_{month} = 29.5\, T_{day}$. For the angle α' in degrees, we have

$$\alpha' = \frac{360}{29.5} = 12.2\,\text{deg}. \tag{A.4}$$

Since $1°$ corresponded to 4 minutes of the Earth's rotation, $12.2°$ corresponds to $12.2 \times 4 = 48.8$ minutes of the Earth's rotation. According to this calculation, the lunar day is longer than the solar day by approximately 49 minutes. This is close to the result of a more precise calculation.

APPENDIX B

Torque on the Earth

N.B. In calculating the torques due to the Sun and the Moon on the Earth in this section, we will ignore the eccentricities of the orbits of both Sun and Moon.

Consider a small part of the Earth of mass m_a, located at $\boldsymbol{r}_a$ from the Earth's center of mass. From Newton's law of gravity the force exerted by the Sun on this mass is

$$\boldsymbol{F}_a = -\frac{GM_s m_a}{|\boldsymbol{r}_a + \boldsymbol{R}_{se}|^3}(\boldsymbol{r}_a + \boldsymbol{R}_{se}). \tag{B.1}$$

The torque on m_a computed about the Earth's center of mass is

$$\boldsymbol{K}_a = \boldsymbol{r}_a \times \boldsymbol{F}_a = -\frac{GM_s m_a}{|\boldsymbol{r}_a + \boldsymbol{R}_{se}|^3}(\boldsymbol{r}_a \times \boldsymbol{R}_{se}). \tag{B.2}$$

Since $|\boldsymbol{r}_a| << |\boldsymbol{R}_{se}|$, we expand $|\boldsymbol{r}_a + \boldsymbol{R}_{se}|^{-3}$ in powers of

$1/R_{se}$, which gives

$$\frac{1}{|\boldsymbol{r}_a + \boldsymbol{R}_{se}|^3} = \frac{1}{R_{se}^3}(1 + 2\frac{\boldsymbol{r}_a \cdot \boldsymbol{R}_{se}}{R_{se}^2} + \frac{r^2}{R_{se}^2})^{-3/2} \sim \frac{1}{R_{se}^3}(1 - 3\frac{\boldsymbol{r}_a \cdot \boldsymbol{R}_{se}}{R_{se}^2} + \ldots). \quad \text{(B.3)}$$

The 1 in the last term of Eq. (B.3) gives zero in the sum over all masses m_a because the origin for $\boldsymbol{r}_a$ is the center of mass of the Earth. The terms denoted ... are higher order in $1/R_{se}$ and can be neglected. We obtain for the total Sun-Earth torque $\boldsymbol{K}_{se}$,

$$\boldsymbol{K}_{se} = \sum_a \boldsymbol{K}_a = 3\frac{GM_s}{R_{se}^5} \sum_a m_a(\boldsymbol{r}_a \times \boldsymbol{R}_{se})(\boldsymbol{r}_a \cdot \boldsymbol{R}_{se}). \quad \text{(B.4)}$$

We can readily evaluate $\boldsymbol{K}_{se}$ when the Earth is at winter solstice. We use a coordinate system with origin at the center of mass of the Earth. The z axis points along the Earth's axis of rotation, so $\hat{\boldsymbol{z}} = \hat{\boldsymbol{n}}$. The x axis is in the same plane as $\hat{\boldsymbol{n}}$ and $\boldsymbol{R}_{se}$ and makes an angle θ_t with $\boldsymbol{R}_{se}$. The y axis is defined by $\hat{\boldsymbol{y}} = \hat{\boldsymbol{z}} \times \hat{\boldsymbol{x}}$. In this coordinate system, we have

$$\boldsymbol{R}_{se} = R_{se}(\cos\theta_t \hat{\boldsymbol{x}} + \sin\theta_t \hat{\boldsymbol{z}}), \quad \text{(B.5)}$$

and

$$\boldsymbol{r}_a = x_a \hat{\boldsymbol{x}} + y_a \hat{\boldsymbol{y}} + z_a \hat{\boldsymbol{z}}. \quad \text{(B.6)}$$

In working out $\sum_a m_a(\boldsymbol{r}_a \times \boldsymbol{R}_{se})(\boldsymbol{r}_a \cdot \boldsymbol{R}_{se})$, since the Earth is symmetric about the z axis, the moment of inertia tensor is diagonal and only terms in $x_a x_a$ and $z_a z_a$ appear in our expression for K. The result for the torque from the Sun when the Earth is at winter solstice is

$$\boldsymbol{K}_{se} = \hat{\boldsymbol{y}}\frac{3GM_s}{R_{se}^3} \sin\theta_t \cos\theta_t \sum_a m_a(z_a z_a - x_a x_a) \quad \text{(B.7)}$$

In the present notation, the moments of inertia C and A are

$$C = \sum_a m_a(x_a x_a + y_a y_a) \tag{B.8}$$

and

$$A = \sum_a m_a(z_a z_a + y_a y_a). \tag{B.9}$$

Using these formulas, we have

$$\boldsymbol{K}_{se} = -\hat{\boldsymbol{y}}\frac{3GM_s}{R_{se}^3}\sin\theta_t\cos\theta_t(C - A). \tag{B.10}$$

Note that $\boldsymbol{K}_{se} = 0$ for $C = A$ (spherical Earth), and that the sense of the torque for any θ_t less than 90° is to twist the Earth toward $\theta_t = 0$.

Now consider the Sun-Earth torque as the Earth moves on its orbit away from the winter solstice. In the coordinate system we are using, the vector $\boldsymbol{R}_{se}$ rotates around the normal to the ecliptic plane. This rotation can be described by an angle $\phi = (2\pi t)/T_{year}$, where T_{year} is 1 year. At an arbitrary value of ϕ, the Sun-Earth torque becomes

$$\boldsymbol{K}_{se} = -\cos^2\phi\hat{\boldsymbol{y}}\frac{3GM_s}{R_{se}^3}\sin\theta_t\cos\theta_t(C - A) + \sin\phi\cos\phi\boldsymbol{K}'_{se}. \tag{B.11}$$

We will not need the expression for $\boldsymbol{K}'_{se}$ since, as we show shortly, it does not contribute to our final expression for the torque. Now the quantity we are seeking, the rate at which $\hat{\boldsymbol{n}}$ precesses about the normal to the ecliptic plane, has a period of many thousands of years. We can smooth out the yearly oscillations of $\boldsymbol{K}_{se}$ by averaging over a period of years much smaller than the period we are seeking, but large enough to replace $\cos^2\phi$ by its average of $1/2$, and $\sin\phi\cos\phi$ by its average of zero. An averaging period as small as 10 years does this to 1% accuracy. Denoting the time-averaged torque by $\bar{\boldsymbol{K}}_{se}$, we have the simple formula

$$\bar{\boldsymbol{K}}_{se} = -\hat{\boldsymbol{y}}\,\frac{3GM_s}{2R_{se}^3}\sin\theta_t\cos\theta_t(C - A). \tag{B.12}$$

The Moon also exerts a torque on the Earth. If we ignore the 5.1° tilt of the Moon's orbit relative to the ecliptic plane normal, the calculation of the Moon-Earth torque is done in the same way as the Sun-Earth torque. The time-averaged total torque on the Earth due to the Sun and Moon is then

$$\bar{\boldsymbol{K}}_{tot} = -\frac{3}{2}\hat{\boldsymbol{y}}\sin\theta_t\cos\theta_t(C - A)(\frac{GM_s}{R_{se}^3} + \frac{GM_m}{R_{me}^3}). \tag{B.13}$$

Since $\hat{\boldsymbol{y}}$ is in the ecliptic plane, so is $\bar{\boldsymbol{K}}_{tot}$. The direction of $\bar{\boldsymbol{K}}_{tot}$ is perpendicular to both the normal to the ecliptic plane, $\hat{\mathbf{e}}$, and the unit vector along the Earth's normal, $\hat{\boldsymbol{n}}$.

APPENDIX C

Rayleigh Waves

As mentioned in Chap. (13), surface waves generated by earthquakes are among the most destructive earthquake effects. In this appendix we will discuss the simplest form of surface wave, the Rayleigh wave, first analyzed in 1885 by John Strutt, Lord Rayleigh. Like Lord Kelvin, Lord Rayleigh was a British scientist known for major investigations in both theory and experimental science. Rayleigh waves generated in an earthquake often travel around the entire Earth several times. However, the frequency-wavelength relation or dispersion relation can be obtained accurately by treating the Earth as locally flat. Being surface waves, Rayleigh waves are damped exponentially upon moving into the Earth. The penetration depth is of the same order of magnitude as the wavelength. The latter ranges from a few kilometers to hundreds of kilometers, so Rayleigh waves often penetrate well into the mantle. However, many of the main features of such waves can be captured by treating the Earth as flat, composed of a single medium with the elastic properties of the

Earth's crust, and this will be done in the following.

Contrast with Bulk Waves There are two important points of contrast between Rayleigh waves and the S and P waves discussed in Sec. (12.2). For both S and P waves, boundary conditions played no role in determining the dispersion relation or, equivalently, the speed of the wave. For Rayleigh waves, as we will see below, boundary conditions at the Earth's surface play a crucial role in determining the dispersion relation. Furthermore, for bulk waves, oscillations along and perpendicular to the direction of propagation are independent. Since they move at different speeds, there is no phase relation between the displacements of S and P waves. On the other hand, one of the most characteristic features of Rayleigh waves is that oscillations parallel to and perpendicular to the direction of propagation move at the same speed and are exactly 90 degrees out of phase.

Meaning of "Longitudinal" and "Transverse" Take the $x-y$ plane parallel to the Earth's surface, with the z axis pointing upward from the Earth's surface, the Earth's surface being at $z=0$. A Rayleigh wave propagates in a direction parallel to the Earth's surface and is damped exponentially as z moves from $z=0$ toward negative values inside the Earth. If the meanings of "longitudinal" and "transverse" are generalized, such a wave can still be broken into longitudinal and transverse parts. A longitudinal displacement $\boldsymbol{u}_L$ is defined as one with no curl, i.e.,

$$\boldsymbol{\nabla} \times \boldsymbol{u}_L = 0, \tag{C.1}$$

while a transverse displacement is defined to be one with no divergence, i.e.,

$$\boldsymbol{\nabla} \cdot \boldsymbol{u}_T = 0. \tag{C.2}$$

The direction of propagation can be any direction in the $x\,y$ plane. For convenience, we take it to be the x direction. The space-time dependence of longitudinal and transverse displacements are then of the form

$$\boldsymbol{u}_L \sim \exp(ikx + \kappa_L z - i\omega t) \text{ and } \boldsymbol{u}_T \sim \exp(ikx + \kappa_T z - i\omega t). \tag{C.3}$$

To satisfy boundary conditions at the Earth's surface ($z = 0$), it is necessary that both terms vary the same way in x and t.

To see how Eqs. (C.1) and (C.2) may be satisfied, apply the gradient operator to the space-time factor of, say, $\boldsymbol{u}_L$. We obtain

$$\boldsymbol{\nabla} \exp(ikx + \kappa_L z - i\omega t) = (ik\hat{\boldsymbol{x}} + \kappa_L \hat{\boldsymbol{z}}) \exp(ikx + \kappa_L z - i\omega t). \tag{C.4}$$

It is easy to see that if we set

$$\boldsymbol{u}_L = A_L(ik\hat{\boldsymbol{x}} + \kappa_L \hat{\boldsymbol{z}}) \exp(ikx + \kappa_L z - i\omega t), \tag{C.5}$$

then $\boldsymbol{\nabla} \times \boldsymbol{u}_L = 0$. Likewise, if we set

$$\boldsymbol{u}_T = A_T(\kappa_T \hat{\boldsymbol{x}} - ik\hat{\boldsymbol{z}}) \exp(ikx + \kappa_T z - i\omega t), \tag{C.6}$$

then $\boldsymbol{\nabla} \cdot \boldsymbol{u}_T = 0$ is satisfied.[1] The full displacement is the sum of $\boldsymbol{u}_L$ and $\boldsymbol{u}_T$,

$$\boldsymbol{u} = \boldsymbol{u}_L + \boldsymbol{u}_T. \tag{C.7}$$

[1] $\boldsymbol{\nabla} \cdot \boldsymbol{u}_T = 0$ is also satisfied if $\boldsymbol{u}_T$ points along $\hat{\boldsymbol{y}}$, but this is needed only for Love waves.

Equation of Motion Both $\boldsymbol{u}_L$ and $\boldsymbol{u}_T$ satisfy their respective wave equations,

$$\frac{\partial^2 \boldsymbol{u}_L}{\partial t^2} = c_L^2 \boldsymbol{\nabla} \cdot \boldsymbol{\nabla} \boldsymbol{u}_L \tag{C.8}$$

and

$$\frac{\partial^2 \boldsymbol{u}_T}{\partial t^2} = c_T^2 \boldsymbol{\nabla} \cdot \boldsymbol{\nabla} \boldsymbol{u}_T. \tag{C.9}$$

Using our forms for $\boldsymbol{u}_L$ and $\boldsymbol{u}_T$, these become

$$\omega^2 \boldsymbol{u}_L = c_L^2 (k^2 - \kappa_L^2) \boldsymbol{u}_L \tag{C.10}$$

and

$$\omega^2 \boldsymbol{u}_T = c_T^2 (k^2 - \kappa_T^2) \boldsymbol{u}_T. \tag{C.11}$$

Combining the last two equations, we have

$$\omega^2 = c_L^2 (k^2 - \kappa_L^2) = c_T^2 (k^2 - \kappa_T^2). \tag{C.12}$$

Note that Eq. (C.12) does not yet determine the velocity of Rayleigh waves, but it does show that κ_L^2 and κ_T^2 are not independent.

Boundary Conditions The pressure and shear stress exerted on the surface of the Earth by the atmosphere are negligible on a geophysical scale. For the purpose of boundary conditions on elastic waves traveling near the surface of the Earth, they may be set to zero. Our boundary condition is then that the pressure and shear stress evaluated by a Rayleigh wave traveling near the Earth's surface must vanish at $z = 0$. Since our wave depends only on x and z, these conditions are

$$\sigma_{xz}(x, 0) = \sigma_{zz}(x, 0) = 0. \tag{C.13}$$

Strain Components The stress components at the surface of the Earth will be expressed by Hooke's law in terms of the strain components u_{xx}, u_{xz}, and u_{zz}. The strain components are easily calculated using the forms given in Eqs. (C.5), (C.6), and (C.7). Since we are at $z = 0$, and the x and t dependence is common to all terms, we will record only the amplitudes of the various strain components. We have

$$u_{xx} = \frac{\partial u_x}{\partial x} = ik(ikA_L + \kappa_T A_T), \quad \text{(C.14)}$$

$$u_{zz} = \frac{\partial u_z}{\partial z} = \kappa_L^2 A_L - ik\kappa_T A_T, \quad \text{(C.15)}$$

$$u_{xz} = \frac{1}{2}(\frac{\partial u_z}{\partial x} + \frac{\partial u_x}{\partial z}) \quad \text{(C.16)}$$

$$= \frac{1}{2}[2ik\kappa_L A_L + (k^2 + \kappa_T^2)A_T].$$

Stress Components The stress tensor is given in terms of the strain tensor by Eq. (12.22). Using that formula, the expression for the shear stress is particularly simple. We have

$$\sigma_{xz} = 2\mu u_{xz}, \quad \text{(C.17)}$$

so the vanishing of σ_{xz} at the Earth's surface is accomplished if the amplitude $u_{xz} = 0$. From Eq. (C.16) this gives

$$2ik\kappa_L A_L + (k^2 + \kappa_T^2)A_T = 0. \quad \text{(C.18)}$$

This equation is of interest, since it determines the ratio of A_L to A_T,

$$\frac{A_L}{A_T} = i\frac{(k^2 + \kappa_T^2)}{2k\kappa_L}. \quad \text{(C.19)}$$

Using this expression for A_L/A_T, it is easy to show that the amplitude ratio u_x/u_z is imaginary; in other words, the displacements in x and z directions are 90° out of phase. This is an important hallmark of Rayleigh waves, clearly seen in seismograph records.

We have still to satisfy $\sigma_{zz} = 0$ at the Earth's surface. Using Eq. (12.22) again, this gives the condition

$$[K(\kappa_L^2 - k^2) + \frac{2\mu}{3}(2\kappa_L^2 + k^2)]A_L - 2i\mu k\kappa_T A_T = 0. \quad \text{(C.20)}$$

Before proceeding, it is useful to simplify the coefficient of A_L in Eq. (C.20). Expressing the ratio $(c_T/c_L)^2$ of Eq. (12.50) in terms of the bulk and shear moduli, we have

$$(\frac{c_T}{c_L})^2 = \frac{\mu}{K + \frac{4\mu}{3}}. \quad \text{(C.21)}$$

Using this and Eq. (C.12), which relates κ_L^2 and κ_T^2, Eq. (C.20) simplifies to

$$\mu(k^2 + \kappa_T^2)A_L - 2i\mu k\kappa_T A_T = 0. \quad \text{(C.22)}$$

At this point, in Eqs. (C.18) and (C.22), we have two homogeneous equations in the two unknowns, A_L and A_T. For a nontrivial solution, the determinant of the coefficients in these two equations must vanish. This gives

$$4k^2\kappa_T\kappa_L - (k^2 + \kappa_T^2)^2 = 0. \quad \text{(C.23)}$$

The Rayleigh Wave Velocity Eq. (C.23) ultimately determines the dispersion relation obeyed by Rayleigh waves. This means that given a frequency ω the wave vector k is determined, or vice versa. To obtain an equation which solely involves ω, k, and elastic constants, κ_L and κ_T must be eliminated. We first square Eq. (C.23) and obtain

$$16k^4\kappa_L^2\kappa_T^2 = (k^2 + \kappa_T^2)^4. \quad \text{(C.24)}$$

Substituting for κ_L^2 and κ_T^2 from Eq. (C.12), we obtain

$$16k^4(k^2 - (\frac{\omega}{c_L})^2)(k^2 - (\frac{\omega}{c_T})^2) = (2k^2 - (\frac{\omega}{c_T})^2)^4. \quad \text{(C.25)}$$

We then introduce the dimensionless variables

$$\zeta \equiv (\frac{\omega}{kc_T})^2, \quad \eta \equiv (\frac{c_T}{c_L})^2. \quad \text{(C.26)}$$

Substituting into Eq. (C.24) and canceling some terms, we arrive at

$$\zeta^3 - 8\zeta^2 + 8(3 - 2\eta)\zeta - 16(1 - \eta) = 0. \quad \text{(C.27)}$$

From the form given for $(c_T/c_L)^2$ in Eq. (12.50), it follows that η is a function of Poisson's ratio, so given a value for Poisson's ratio, a solution of Eq. (C.27 determines the dispersion relation for Rayleigh waves. There is only one real root of this equation so the dispersion relation is unique. Defining the velocity of a Rayleigh wave as $c_R = \omega/k$, we have

$$c_R = c_T\sqrt{\zeta}. \quad \text{(C.28)}$$

By examining the solution of the cubic equation, it is found that ζ is always less than 1 as the Poisson ratio varies from 0 to 1/2, so the velocity of Rayleigh waves is always smaller than c_T. For the case where the Earth's crust is a "Poisson solid" with a Poisson ratio of 0.25, the root of Eq. (C.27) is $2 - 2/\sqrt{3}$, so

$$c_R = c_T[2 - \frac{2}{\sqrt{3}}]^{1/2} \sim 0.92c_T.$$

As an example, take $c_T = 3\,\mathrm{km/s}$, so $c_R \sim 2.76\,\mathrm{km/s}$. For a period of the Rayleigh wave of 30 s, the wavelength of the Rayleigh wave is

$$\lambda_R = c_R T_R = 2.76 \times 30 \sim 83\,\mathrm{km}.$$

Penetration Depths The amplitude of a Rayleigh wave falls exponentially with increasing depth. There are of course two terms, one varying as $\exp(\kappa_L z)$ and the other as $\exp(\kappa_T z)$. We may define corresponding penetration depths,

$$a_L = \frac{1}{\kappa_L} \text{ and } a_T = \frac{1}{\kappa_T}. \tag{C.29}$$

These define the depths at which the longitudinal and transverse amplitudes fall to $1/e$ of their surface values. From Eq. (C.12), we have

$$\kappa_T^2 = k^2(1 - (\frac{c_R}{c_T})^2). \tag{C.30}$$

Solving for a_T, setting $k = 2\pi/\lambda_R$, and continuing to assume a Poisson ratio of 0.25, we have

$$a_T = \frac{\lambda_R}{2\pi\sqrt{1-(c_R/c_T)^2}} = 0.405\,\lambda_R. \tag{C.31}$$

The corresponding result for a_L is

$$a_L = 0.241\,\lambda_R. \tag{C.32}$$

For the example of the previous section which had $\lambda_R \sim$ 83 km, the penetration depths are at most 40 km, so for such waves the displacement is still significant in the upper region of the mantle.

Bibliography

Ahrens, T. J., ed. 1995. *Global Earth Physics: A Handbook of Physical Constants.* Washington, DC: American Geophysical Union.

Barnett, S. J. 1935. "Gyromagnetic and Electron-Inertia Effects." *Rev. Mod. Phys.* 7:129–166.

Bleeker, J. A., J. Geiss, and M. Huber. 2001. *The Century of Space Science.* Berlin: Springer.

Brush, Stephen. 1996. *A History of Modern Planetary Physics: Nebulous Earth.* London: Cambridge University Press.

Callen, H. B. 1960. *Thermodynamics.* New York: John Wiley & Sons.

Campbell, Wallace. 2003. *Introduction to Geomagnetic Fields.* London, Cambridge University Press.

Carslaw, H. S., and J. C. Jaeger. 1959. *Conduction of Heat in Solids.* Oxford: Oxford University Press.

Catling, D., and K. Zahnle. 2009. "The Planetary Air Leak." *Scientific American* 300:36–43.

Chalmers, J. Alan. 1967. *Atmospheric Electricity.* Oxford: Pergamon Press.

Claude, Alexis, and John Colson. 1737. "An Inquiry concerning the Figure of Such Planets as Revolve about an Axis." *Phil. Trans.* 40:277–306.

Cook, A. H. 1973. *Physics of the Earth and Planets.* New York, John Wiley and Sons.

EarthAlabama. 2017. http://earthalabama.com/energy.html.

Emanuel, K. 1986. "An Air-Sea Interaction Theory for Tropical Cyclones. Part I: Steady State Maintenence." *J. Atm. Sci.* 43:585–605.

———. 1991. "The Theory of Hurricanes." *Annu. Rev. Fluid Mech.* 23:179–196.

Fermi, Enrico. 1956. *Thermodynamics.* New York: Dover.

Feynman, Richard, R. B. Leighton, and M. Sands. 1966. *The Feynman Lectures on Physics, Vol. II.* Reading, MA: Addison-Wesley.

Fowler, C. M. R. 1990. *The Solid Earth.* Cambridge: Cambridge University Press.

Galison, Peter. 1987. *How Experiments End.* Chicago: University of Chicago Press.

Goldstein, H. 1999. *Classical Mechanics, Second Edition.* New York: Addison-Wesley.

Gross, Peter G. 1974. "A Lower Limit to Jeans' Escape Rate." *Mon. Not. R. Astr. Soc.* 167:215–219.

Holzworth, Robert H. 1991. "Conductivity and Electric Field Variations with Altitude in the Stratosphere." *J. Geophys. Res. Atmos.* 96:12857–12864.

Huang, Kerson. 1987. *Statistical Mechanics.* New York: John Wiley & Sons.

Jackson, John D. 1999. *Classical Electrodynamics.* New York: John Wiley and Sons.

Jackson, Patrick Wyse. 2006. *Episodes in the Search for the Age of the Earth.* Cambridge: Cambridge University Press.

Jeans, J. 1916. *Dynamical Theory of Gases.* London, Cambridge University Press.

Kanamori, H. 1977. "The Energy Release in Great Earthquakes." *J. Geophys. Res.* 82:2981–2987.

Kinsman, Blair. 1965. *Wind Waves.* Englewood Cliffs, NJ, Prentice-Hall.

Krall, N. A., and A. W. Trivelpiece. 1973. *Principles of Plasma Physics.* New York: McGraw-Hill.

Landau, L. D., and E. M. Lifshitz. 1959. *Fluid Mechanics.* New York: Pergamon Press.

———. 1960. *Electrodynamics of Continuous Media.* New York: Pergamon Press.

———. 1976. *Mechanics, Third Edition.* New York: Pergamon Press.

———. 1980. *Statistical Physics, Third Edition, Part 1.* New York: Pergamon Press.

———. 1986. *Theory of Elasticity, Third Edition.* New York: Pergamon Press.

Loeb, L. B. 1958. *Space Electrification.* Berlin: Springer.

Love, A. E. H. 1911. *Some Problems of Geodynamics.* London: Cambridge University Press.

Mandl, F. 1971. *Statistical Physics.* New York: John Wiley & Sons.

Marshall, John, and R. Alan Plumb. 2008. *Atmosphere, Ocean, and Climate Dynamics: An Introductory Text.* London: Elsevier Academic Press.

Masters, G., and S. Constable. UCSD Lecture Notes, sio103, Chapter 3. http://igppweb.ucsd.edu/guy.

Nerem, R. S. 1994. "Gravity Model Development for TOPEXPOSEIDON; Joint Gravity Models 1 and 2." *J. Geophys. Res. Oceans* 99:24, 421–424, 447.

Ness, N. F., J. C. Harrison, and L. B. Slichter. 1961. "Observations of the Free Oscillations of the Earth." *J. Geophys. Res.* 66:621–629.

Pavlis, N. K., S. A. Holmes, S. C. Kenyon, and J. K. Factor. 2012. "The Development and Evaluation of the Earth Gravitational Model 2008 (EGM2008)." *J. Geophys. Res.* 117:B04406.

Press, F., and R. Siever. 1985. *The Earth.* New York: W. H. Freeman.

Qin, J., and L. Waldrop. 2016. "Non-thermal Hydrogen Atoms in the Terrestrial Upper Atmosphere." *Nat. Comm.* 7:13655.

Rawlinson, N. Australian National University Lecture Notes. http://rses.anu.edu.au/nick.

Rayleigh, Lord. 1885. "On Waves Propagated along the Plane Surface of an Elastic Solid." *Proc. London Math. Soc.* s1-17:4–11.

Richter, Charles F. 1958. *Elementary Seismology.* San Francisco: W. H. Freeman.

Rossby, C. G. 1939. "Relation between Variations in the Intensity of the Zonal Circulation of the Atmosphere and the Displacements of the Semi-permanent Centers of Action." *J. Mar. Res.* 2:38.

Schubert, G. 2009. *Treatise on Geophysics, Volume 1, Seismology and the Structure of the Earth.* New York: Elsevier.

Shreve, E. L. 1970. "Theoretical Derivation of Atmospheric Ion Concentrations, Conductivity, Space Charge Density, Electric Field and Generation Rate from 0 to 60 km." *J. Atmos. Sci.* 27:1186–1194.

Smith, R. Lectures on the Dynamics and Thermodynamics of Tropical Cylones. http://en.meteo.physik.uni-muenchen.de.

Stacey, Frank D. 1969. *Physics of the Earth.* New York: John Wiley and Sons.

Stein, Seth, and Michael Wysession. 2003. *An Introduction to Seismology, Earthquakes, and Earth Structure.* Oxford: Blackwell.

Stoneley, Robert. 1961. "The Oscillations of the Earth." *Phys. Chem. Earth* 4:239–250.

Van Boxel, John. 1998. Numerical Model for the Fall Speed

of Raindrops. University of Amsterdam I.C.E. Report.

Volland, Hans. 1984. *Atmospheric Electrodynamics.* Berlin, Springer.

Wahlin, Lars. 1989. *Atmospheric Electrostatics.* New York: Research Study Press, John Wiley & Sons.

———. 1994. "Elements of Fair Weather Electricity." *J. Geophys. Res.* 99:10767–10772.

Wilson, A. H. 1957. *Thermodynamics and Statistical Mechanics.* Cambridge: Cambridge University Press.

Yoder, C. F. 1995. In Ahrens 1995, 1–31.

Index

albedo, 112
alpha particle, 78
atmosphere
  adiabatic, 47
  composition of, 63
  effects of gravity on, 44
  isothermal, 45
  Jeans escape velocity, 71
  lapse rate, 49
  layers of, 239
  upper, 65
aurora borealis, 234

basalt, 203
black body radiation
  Planck and, 110
  Stefan-Boltzmann, 110
  Wien's law, 111

capillary action, 86
carnot cycle
  gas, 39
  in hurricane , 103
centrifugal gravity effects, 13
chlorofluorocarbons, 64
Clairaut's formula, 17
Coriolis effects
  jet streams, 102
  wind circulation, 98

day
  lunar, 265
  sidereal, 8, 263
  solar, 8, 263

Earth
  breather mode, 189
  central pressure, 21
  Chandler wobble, 32
  composition, 199, 203
  core, 179
  crust, 108
  density of, 9
  equinox precession, 25
  Euler's nutation, 32
  flattening parameter, 9
  gravitational data, 8
  gravitational potential, 23
  magnetic moment, 224
  mantle, 179, 183, 184, 193
  moment of inertia, 22
  negative charge on , 239
  solar and sidereal day, 8
  torque on, 26, 267
  volume of, 9
earthquakes
  energy release, 159
  Mercalli scale, 157
  San Francisco, 159
  seismic moment, 158
  seismographs, 163
ecliptic plane, 24
elasticity theory
  bulk modulus, 174

  - Hooke's law, 171
  - Newton's equation, 175
  - Poisson's ratio, 172
  - shear modulus, 174
  - strain, 165
  - stress, 170
  - Young's modulus, 172
- electrosphere
  - earthward current in, 238
  - electric conductivity of, 243
  - ions in, 240
  - sources of ions, 242
- electrospraying, 249
- exosphere, 76

- fluid flow
  - Euler's equation, 132
  - Reynold's number, 89

- geographical coordinates, 10
- geological eras
  - length of, 209
- geomagnetic field
  - anomalies, 226
  - Barnett effect, 228
  - De Haas-Einstein effect, 228
  - dipole representation, 235
  - geodynamo, 230
  - reversals, 232
  - source of, 230
- glacier flow, 58
- granite, 203
- gravitational equipotentials, 14

- heat equation, 122

- ideal gas
  - adiabatic expansion, 39
  - internal energy, 42
  - specific heats, 42
- ionosphere
  - electron plasma in, 252
  - properties of, 251
  - radio wave reflection, 255
  - transverse waves in, 255

- liquid drops
  - in electric field, 247
  - inside pressure, 84
  - radius, 85
  - terminal velocity, 89

- mantle, 108

- nonspherical gravity effects, 10

- radioactivity of Earth
  - effect on Earth age, 201
  - important isotopes in, 203
  - uranium to lead decay, 206

- San Francisco earthquake, 157
- seismic waves
  - longitudinal, 178
  - Love, 187
  - Rayleigh, 187, 271
  - transverse, 178
- solar luminosity, 109
- solar wind, 78

- tides
  - lunar, 143, 151
  - on lakes and seas, 154
  - solar, 153
  - time dependence, 150